PSILOCYBIN
AND MAGIC MUSHROOMS FOR BEGINNERS

THE ULTIMATE GUIDE TO
PSYCHEDELIC PSILOCYBIN MUSHROOMS –
HOW TO GROW AND CULTIVATE THEM,
USE THEM FOR SPIRITUAL HEALING,
THEIR HISTORY, BENEFITS AND MORE!

JEFFREY MACMILLMAN

DISCLAIMER

The information in this audiobook is intended for educational purposes only. The authors and the publisher do not advocate breaking the law. Although the authors and the publisher have made every effort to ensure that the information inside this audiobook was correct, accurate, up to date and reliable at the time of publishing, the authors and publishers do not assume and hereby disclaim any liability to any party for any loss, damage, or disruption caused by errors or omissions, whether such errors or omissions result from negligence, accident, or any other cause. By listening to this audiobook, the listener agrees that under no circumstances are the authors or the publisher is responsible for any losses, direct or indirect, which are incurred as a result of the use of the information contained within this document.

Please consult a licensed professional before attempting any of the techniques outlined in this audiobook.

★ ★ ★ ★ ★

Thank you for getting our book!

If you enjoy reading and find it useful we would greatly appreciate your review on Amazon.

Just head on over to this book's Amazon page and click "Write a customer review".

We read each and every one of them. Thanks!

★ ★ ★ ★ ★

CONTENTS

INTRODUCTION

THE MAGIC OF MUSHROOMS

"I learned more about psychology in the five hours after taking these mushrooms than in the preceding 15 years of studying and doing research in psychology."
- TIMOTHY LEARY

Fungi, one of the six great kingdoms of life as we know it, consists of a vital conglomeration of organisms who have shaped the history and ecology of planet Earth for approximately 2.4 billion years.

They are the decomposers-- the custodians of the planet-- cleaning up behind the other organisms who thoughtlessly litter the terrain with their unsightly remains of death, and in so doing, provide the substrate for the furtherance of life.

Though each member of this kingdom is in itself immensely impressive and compelling, we will, for the time being, dismiss those fungi which serve both to plague or to improve our health and our foodstuffs: the blue and green molds which cool our fevers, flavor our cheeses, and decay our fruits; the abundant yeasts which bubble in our brews, lift up our breads, and pattern our palates with painful thrush; the meadow mushrooms, the shiitakes, the truffles, each a treasured morsel which serve to nourish us and expand our diets with a delightful variety of textures and flavors.

These former are magical in their own sense, but our subject is of a different variety altogether; a contentious member of this expansive family, one that has elevated lives as well as cast them down to confused ruination.

We seek today a toadstool, a rather innocuous and inconspicuous entity, which holds within its many-threaded structure a vast and awesome secret. We are speaking of psilocybin mushrooms.

WHAT ARE PSILOCYBIN MUSHROOMS?

Psilocybin mushrooms are, in simplest terms, mushrooms that contain the psychoactive chemicals psilocybin, and to a lesser extent, psilocin. We will talk in much greater length about these chemicals, but first, let's jump out of the psilocybin umbrella and into the wider dimension of the term "magic mushroom."

Most times, people use the term *magic mushroom* to refer specifically to psilocybin-containing species, but there exists the possibility that they are referring to a different type of psychoactive mushrooms in the genus *Amanita*. Though of far less common use than psilocybin mushrooms, certain species of the *Amanita* genus contain a cocktail of psychoactive chemicals, including muscimol, muscarine, and ibotenic acid. The mind-altering effects of these chemicals differ greatly from psilocybin but have similarly been used ceremonially and recreationally for many years.

We will return to *Amanitas* at a later point in this book, but the majority of our focus will be on the much more popular psilocybin-containing fungi.

There are at least 250 different species of mushroom known to contain psilocybin, distributed among two phyla and approximately 22 genera.[1] Amazingly, they can be found on every continent with the exception of Antarctica.[2]

Among the genera containing "magic" species are *Copelandia*, *Gymnopilus*, *Inocybe*, *Panaeolus*, *Pholiotina*, *Pluteus*, and *Psilocybe*, with the majority of known species found in the *Psilocybe* genus. *Psilocybe cubensis* is perhaps the most commonly known, but it is by no means the only magic mushroom in common use.

Most, but not all *Psilocybe* species are known to be neurotropic. At this point, you might take a guess that *psilocybe* is a Latin word translating to something along the lines of "crazy brain mushroom" or "fungus which bends the mind," but the genus name actually comes from the Greek words ψιλός (*psilo*), meaning bald, bare, or smooth, and κύβη (kúbē), meaning head. This name is in reference to the smooth, unblemished appearance of these mushrooms' caps.

What Do They Look Like?

In general, *Psilocybe* mushrooms are small- to medium-sized fungi that range in color from brown to tan to a kind of muddy yellow. Famously, they are said to exhibit blue bruising when pressed, squeezed, or scratched.

Now, before you go eating the next little brown mushroom you find, it's important to know that there is a plethora of similar-looking species, some of which are unpleasantly toxic. You must also know that there are plenty of blue-staining mushrooms which are not psychedelic. Being small and brown and staining blue <u>are not</u> sufficient-enough traits to identify *Psilocybe* species with. At this point, we need to tell you that <u>eating wild mushrooms can be dangerous and you should only do this with absolute certainty of what you are picking.</u> Please, for your own sake, do not pick and eat wild mushrooms without proper education on the subject.

Are Psilocybin Mushrooms Safe?

We will avoid the overuse of the words *drug* and *substance* throughout this book but find it important that we now clarify our meaning and purpose of using those terms. The word *drug* has, due to the legal restrictions placed on certain molecules and compounds by various governmental agencies, received an overall negative connotation in popular culture. *Drug*, when used outside strictly medical applications, brings to mind dangerous chemicals with high incidences of addiction and life-ruining

10

consequences. *Substance,* in similar contexts, holds much the same association.

In this book, the authors have chosen to opt for a more literal use of language and to spurn the undercurrents of meaning cultural understanding has attached to otherwise useful locutions. When we use the word *drug,* we mean not a recreational substance that harms the body and degrades society. We do not imply an illicit chemical compound, nor a modern prescription medication, nor a narcotic, stimulant, depressant, or any other meaning outside the simplest literal definition of the word. We will use *drug* as it is defined in pharmacology; that is, a chemical compound which produces a physiological effect when consumed. *Drugs* do not exist in a strict dichotomy of good and bad. They simply are.

In this sense, psilocybin and psilocin are indeed *drugs,* just as are acetaminophen, heroin, and sugar. They are also vastly misunderstood drugs which have been unfairly demonized in modern times by authoritative bodies with unclear intent.

First, we will set a few things straight. For starters, psilocybin is not physically or psychologically addictive, nor does it have a high potential for compulsive use.[3] Users of psilocybin do not become dependent on it in the same way that users of opiates become dependent on those drugs. On the contrary, it is currently being researched and showing great promise as a treatment for addiction.[4]

Furthermore, psilocybin is a relatively safe drug in terms of life-threatening reactions. Indeed, there have only been two recorded deaths attributed to psilocybin mushroom ingestion, neither of which have made a conclusive link between the fatality and the consumption of the mushrooms. The LD50, or lethal dose, of psilocybin has been determined to be 280 mg/kg in rats,[5] compared to an LD50 of 50 mg/kg for nicotine, 110 mg/kg for alcohol, and 37 mg/kg of Vitamin D3. To put this in perspective, by these calculations a 180 lb person would need to eat roughly 42 pounds of pure psilocybin to experience a fatal reaction.

There are, as with most things, certain dangers to psilocybin use, such as a possible link between hallucinogens and subsequent psychiatric disorders. There is also an inherent risk in impairing one's consciousness, which accounts for the majority of psilocybin-related injury. A chapter of this book is devoted to safe psilocybin use in which we address these issues in more depth. All in all, psilocybin is one of the least harmful, and perhaps most beneficial drugs known to humankind.

What Do Magic Mushrooms Do?

Psilocybin mushrooms are often called *hallucinogenic*, but the authors are not fond of this term and will eschew it for classifications we find more appropriate. *Hallucinogenic*, meaning "causes hallucinations" is a poor descriptor that we feel unjustly depreciates the effects of magic mushrooms. They do much more than cause so-called hallucinations. For this type of drug, we prefer the terms *psychoactive* and *psychotropic*, which simply indicate that they affect one's mental state. Even more appropriate might be the term *entheogen*, a label coined in 1979 to refer to drugs that induce a spiritual or religious experience. However, since not everyone may report a religious experience with magic mushroom use, we will for the majority of this book use the simple *psychotropic, psychedelic,* and *psychoactive* designations.

These psychotropic mushrooms have a variety of effects on their users that vary widely depending on the individual and the conditions under which they are consumed. Their effects can range from overwhelmingly positive to mildly disastrous, though their beneficial results are reported with greater prevalence and with more enduring repercussions.

Experiences with magic mushrooms are popularly called trips, as in "I had a mushroom trip," or "I was tripping yesterday." Surprisingly, this term was coined by U.S. Army scientists during their experiments with LSD in the 1950s.[6] After an intense experience with a psychotropic drug such as mushrooms, you will probably understand why the word *trip* is a fitting description. Essentially, these substances cause users to excurse on a vast "journey of the mind."

Typically, the onset of effects (the beginning of the trip) begins between thirty minutes to two hours after ingestion, and the whole experience lasts between four to nine hours.[7] Depending on the amount consumed, positive effects range from giddiness and heightened emotional responses at low doses, to intense feelings of euphoria and spiritual awakening with high doses. Furthermore, different varieties may have different effects, as evidenced by the Mazatec's valuing of certain species over others, yet not much empirical research has been done regarding these distinctions.

Magic mushrooms are not exactly prized for their physical effects, which run from benign pupil dilation to more consequential tremors, nausea, and vomiting. While few serious side effects are usually reported, users must take necessary precautions in regards to their surroundings before ingestion, as lack of coordination is somewhat commonly reported. It is also important that you educate yourself about any preexisting contraindications that may preclude you from safe use.

Doses less than one gram may cause a slight feeling of coldness, mild gas and nausea, stomach cramps, and pupil dilation. Medium doses between one and two and one-half grams can cause the same effects of lower doses and further extend to greater nausea and stomach discomfort, a rise or drop in blood pressure, changes in heart rate, and changes in stretch reflex. The physical effects of doses over two and one-half grams may involve all the former mentioned and can include irregular heart rhythm, tremors, and vomiting.

While these somatic reactions sound unpleasant, they are usually mild in nature and greatly overshadowed by the positive psychological effects. Psychological effects of mushrooms vary greatly at different doses. Low doses involve a heightening of mood, giddiness, joy, and feelings of peace. Those who take very small doses often claim a boost in creative thinking, a calming of anxiety, and an elimination of depressive thoughts. High doses are much more profound. Depending on the mindset of the user, high doses can trigger religious experiences, the sense of being transported to another dimension, and perceived psychic and telepathic abilities. Those with a more secular worldview may simply claim a mundanely interesting psychological experience.[8]

The Many Names of Magic Mushrooms

People around the world call psilocybin mushrooms by many different monikers. In the English language, we, of course, have *magic mushrooms*, but other slang words include *shrooms, boomers, zoomers, caps, mushies, funny mushrooms, goombas/goomies,* and even *pizza toppings*. Among their many epithets in Mexican Spanish are *derrumbes* (landslides), *pajaritos* (little birds), *hongitos* (little mushrooms), *hongos mágicos* (magic mushrooms), or *hongos sagrados* (sacred mushrooms). Traditional names in Mexico from the natives' languages are *apipiltzin, atkad, di nizé taaya, shi thó,* and the Aztec's *teotlaquilnanácatl*. Throughout Spain, they may be called *hongos* (fungi) or *setas* (mushrooms). The Irish use the term *muisiriún mire*, meaning mushroom of madness. In Italian slang, psilocybin mushrooms share the name of a mushroom-shaped pasta, *funghetti*. Norwegians call them *fleinsopp*, the Dutch call them *paddos*, Poles call them *psyfki*, and the French call them *champi*.

These names can allude to their psychoactive effects, their appearance, or their growing conditions, and can carry both literal or figurative meaning. A whole book could be dedicated to the various names of magic mushrooms and how those names were earned, but the point we are making is that they are known and used throughout the world, prevalent enough to have region-specific handles. Despite these diverse nicknames, the most common term for these many species of psychotropic fungi is, in variegated translations, *magic mushrooms*.

OCCULT ORIGINS:
THE HIDDEN PAST OF MUSHROOM MAGIC

There is a thick and obscuring veil covering the vast majority of humanity's time on Earth, a fact which the authors have always regarded with tragic disconsolation. Our greatest fascinations lie in our prehistory, the stories of our ancestors now lost to a sea of myth, legend, and speculation.

With a great hunger for facts, we have remained unsatisfied with historical hypotheses and postulations that are carried by no more evidence than fables and fairy tales.

Such is the situation we encounter when looking back to the beginning interactions between humans and psychoactive mushrooms. We have an agonizingly short proven record of their use, the first mention dating to 1569, found in *Historia general de las Cosas de Nueva España* by the Franciscan friar Bernardino de Sahagún.[9] The next known mentions are found in the works of Spanish naturalist Francisco Hernández de Toledo, written after his scientific expedition to catalog the flora and fauna of the "New

World" in the year 1570.[10] Though it is doubtful the Aztecs began using magic mushrooms the very year of these Spaniards' arrival, we have no concrete evidence of any use prior. Any writings regarding ceremonial or spiritual mushroom use before this time have been lost to history or are otherwise so vague in their mysterious references to the magical substances used that we can not be sure if they speak of mushrooms at all.

There are, however, many references to mushrooms in other artistic historical media, such as carvings, paintings, and oral histories, that precede the 1500s by several millennia. 12,000-year-old rock art in Australia and Tanzania depicts mushroom-headed beings that suggest a possible shamanistic use of psilocybin-containing species.[11] A 9,000 BCE mural in Algeria displays what may possibly be the psychoactive species *Psilocybe mairei*.[12] Then in Spain, there is the Selva Pascual cave mural, containing an orderly row of thirteen mushrooms thought to be *Psilocybe hispanica*.[13]

In Guatemala, stone carvings of anthropomorphic mushrooms have been discovered which date to between 2,500 and 3,000 years ago.[14] Later in Mayan history came the *Codex Vindobonensis Mexicanus I*, a 14th-century Mayan text which contains illustrations of the Mayan god of the underworld holding clusters of fungi.[15]

Prehistoric humans seemed to have quite a fixation on mushrooms, dedicating the time, energy, and materials to their iconography for millennia. What are we to make of this? Theories abound, and none are completely satisfying in their ultimate unprovability, but it suggests through both repetition and prominence the idea that mushrooms were rather significant to our ancient ancestors.

CULTURAL AND CONTEMPORARY BELIEFS ON THEIR SIGNIFICANCE

What is it about these mushrooms that humans have found so significant for such a long amount of time? There is, at face value, something extraordinary about the ingestion of an inconspicuous fungus having such a profound effect on the consciousness, but is there a deeper meaning? We will not take a look at the mythical and spiritual beliefs of various mushroom-using world cultures.

Traditional Aztec Beliefs

Through the destruction of the Spanish conquistadors and subsequent Catholic clergy, our knowledge about the belief systems of the Mayan and Aztec peoples is extremely limited. What little we know comes from a few clerical descriptions with strong negative biases and the more tolerant accounts of Bernardino de Sahagún.

We know that certain psilocybe species were called *teonanácatl* in the Aztec language. This word has often been translated as "God's flesh," but recent research by Guzmán throws this interpretation into question.[16] Following along the lines of this translation, it has been assumed that Aztecs associated psychotropic mushrooms with the feathered-serpent god Qetzocoatl, believing that their ingestion imbued deistic knowledge upon the users. The use of these mushrooms was restricted to priests in ceremonial settings, though one account describes their distribution to a large group of subjects witnessing the coronation of the emperor Montezuma I.[17] There is little we can say with certainty about their traditional Meso-american use, and we will end with that as to not be yet additional Westerners offering misguided assumptions.

Shamanic Use of *Amanita muscaria*

Perhaps due to the inherent secrecy of sorcery and rituals, we can go back only so far in our quest for knowledge of magic mushroom use throughout European history. In the Western world, there are no documented uses of psilocybin mushrooms in either recreational or ceremonial use until their popularization in the 1950s. There are, however, older accounts of the fly agaric mushroom, taxonomically known as *Amanita muscaria.*

Fly agaric has been in use by various Siberian peoples since at least the 17th-century and perhaps much longer. The first account of their use comes to us from a Polish prisoner of war, Kamiensky Dluzyk,[18] in a 1658 journal entry simply describing this fungus' intoxicating effects. Later explorers to the region of eastern Siberia documented not only the mushroom's use, but also mythologies and spiritual beliefs.

The Russian ethnographer Waldemar Jochelson was the first to produce a written account of the mythical origins of *Amanita muscaria*, writing of the Koryak people's legend:

> Once - say the Koriaki - Grande-Corvo had caught a whale, and was unable to take it to his home in the sea [...] Grande-Corvo turned to Existence asking for help. The divinity said to him: "go to a flat place near the sea: there you will find fluffy white stalks with spotted hats. These are the *wa'paq* spirits. Eat some and these will help you. " Grande-Corvo went to the place indicated. Then the Supreme Being spat on the earth and agaric appeared from his saliva. Grande-Corvo found the mushroom, ate it and began to feel joyful. He began to dance. The muscarius agaric said to him, "How is it possible that you, so strong, cannot lift the bag?" - "You're right - said Grande-Corvo - I'm a strong man. I will go and lift the travel bag ". He went, immediately lifted the bag and sent the whale home. Then the agaric showed him how the whale was going to the sea and how it was returning to its companions. Then Great-Crow said "Let the agaric remain on earth and let my children see what it will show them."[19]

This legend gives credence to the belief that ingestion of *Amanita muscaria* can imbue the user with great physical strength and vitality. Jochelson goes on to explain that the Koryak believe that the mushroom will divulge to its users the mysteries of their ailments and afflictions, as well as give commands that direct the actions of the users.

Enderli, in 1903, wrote that the Koryak used the fly agaric to see the future. Before eating it, they were said to speak incantations above the fungus, which would then reveal their future through dreams as they slept.[20]

Of the Mansi and Ostyak peoples, the Finnish explorer Kustaa Karjalainen wrote in 1927 that the fly agaric can provide the user with desired knowledge and information once in a state of trance.[21]

In general, the beliefs of the powers obtained by *Amanita muscaria* held by various Siberian peoples range from communication with spirits and souls of the dead to shamanistic journeys in which the practitioners own soul is said to be able to leave the body and travel to other worlds or dimensions. Purposes for use also include medicinal treatment of diseases, psychic foresight and personal insight, summoning of spirits to be of help in both worldly and spiritual matters, and for increasing general courage.[22]

It is interesting to note that the modern-day Santa Claus has many similarities with the myth and magic of *Amanita muscaria*. For one, the red and white mushroom has obvious parallels with Santa's usual attire. Fly agarics, once gathered from underneath their pine tree habitats into hide sacks, were in some villages distributed as gifts. Furthermore, several of these explorers of the Siberian region noted that reindeer, Santa's preferred means of transportation, would excitedly consume both the mushrooms and the urine of people who had recently eaten the mushrooms. Some of the native Siberian peoples, whose economy was largely based around the trade of reindeer pelts and meat, were known to become intoxicated after eating the flesh of reindeer which were killed shortly after consuming the agarics. Among other coincidences is the structure of the Koryak people's traditional homes, which, according to the 17th-century writings of Filip Johann von Strahlenberg,[23] were entered through a hole in the roof, much like the patron saint of Christmas is said to enter through chimneys. There is debate as to whether or not Santa's origins have any basis in the shamanic and recreational use of fly agaric mushrooms, but the authors appreciate the coincidences and think there may be something to it.

Mazatec and Psilocybin Spirituality

Although the internet buzzes with supposed beliefs held by Mazatec people concerning the use of magic mushrooms, very little is actually known. Shamans in many cultures tend to guard the secrets of their practices, and it is no different with the Mazatec people, who, due to the persecution their ancestors faced, have good reason to shelter their esotericisms.

What has been gathered by Westerners regarding the Mazatec belief system surrounding psilocybin use amounts to a few key elements.[24] Foremost, mushrooms are sacred and divine, sometimes referred to as "Christ's blood" and said to grow from spots where the blood of Christ has fallen on Earth. Secondly, magic mushrooms should only be used for healing purposes. They are often called "little saints" and held to be a connection to or, in fact, be themselves spirits that connect the healer and patient to an ultimate source of knowledge. Furthermore, mushrooms are said to communicate to the user directly and through spoken language. They may disclose the cause and remedy of illnesses, reveal the location of a lost item or person, or impart special knowledge to the user.

Perhaps our best source of information regarding Mazatec mushroom use is the curandera Maria Sabina. In the book *Maria Sabina: Selections*,[25] Alvaro Estrada presented a translation of much of Sabina's oral biography. She recounts her calling to become a curandera and offers her belief that in taking the "little saints," she is able to communicate with greater spirits and be in closer communion with God.

Hippies and Modern Metaphysics

With the start of the counterculture and hippie movements of the 1960s, psychedelic drug use boomed into popularity. People like Timothy Leary and Richard Alpert were vocal advocates during this time, encouraging the use of psilocybin and other psychotropics as a means to advance toward enlightenment at the individual and societal levels. Much of the spirituality of the 1960s hippie movement was influenced and based on beliefs held in the Hindu and Buddhist religions. The most prevalent of these concepts include the beliefs that souls are eternal, all existence is a connected whole, peaceful coexistence is a high virtue, and self-realization is of the utmost importance to each individual.

In a 1966 speech, Leary gave a concise summation of the popular psychedelic spiritual views of the time:

"Like every great religion, we seek to find the divinity within and to express this revelation in a life of glorification and the worship of God. These ancient goals we define in the metaphor of the present—turn on, tune in, drop out."[26]

Leary gave a later explanation of the phrase "turn on, tune in, drop out" in his 1983 autobiography:

"'Turn on' meant go within to activate your neural and genetic equipment. Become sensitive to the many and various levels of consciousness and the specific triggers engaging them. Drugs were one way to accomplish this end. 'Tune in' meant interact harmoniously with the world around you—externalize, materialize, express your new internal perspectives. 'Drop out' suggested an active, selective, graceful process of detachment from involuntary or unconscious commitments. 'Drop Out' meant self-reliance, a discovery of one's singularity, a commitment to mobility, choice, and change. Unhappily, my explanations of this sequence of personal development are often misinterpreted to mean 'Get stoned and abandon all constructive activity.'"[27]

Of course, beliefs are subjective and vary greatly from individual to individual, and without a long history of cultural practice and understanding, the psychedelic spirituality of the first Western psilocybin users was inconsistent and oftentimes confused and contradictory.

In more recent times, the New Age movement has characterized much of the West's beliefs in the area of psychedelic spiritual use. New Age spirituality is broad and largely nonspecific, drawing heavily from various global religions such as Hinduism, Buddhism, neopaganism, and Ancient Egyptian mythology.[28] There is still the strong undercurrent of the 1960s in the views that interconnectivity, connection with nature, and inner self-exploration are vitally important aspects of life, and that the use of psilocybin mushrooms is one way to facilitate these practices.[29]

Further spiritual beliefs among some modern psilocybin users, though certainly not all, include the ability of magic mushrooms and other psychedelics to elevate one's consciousness to a higher level, to allow the user to travel to other dimensions of reality, to enable the user to communicate with spirits or extraterrestrial beings, and the ability to see the inherent "truth of reality."

Doubtless, there are many who take psilocybin mushrooms simply for fun and recreation and still others who have vastly different beliefs than any we have mentioned here. Spirituality is, on the whole, a completely subjective topic, and to define it with any certainty for a group as varied as magic mushroom users is an altogether impossible task.

SALUE FOR THE SPIRIT

NATURE AS THE PRIMORDIAL CURE

Life and nature are inseparable parts of the same whole, and the authors believe that the natural world provides for the needs of each living creature. The planet is a closed, self-sustaining system, where each part fills its own helpful little niche.

Our oldest ancestors understood this and found the remedies to their ailments in the multifarious forms of flora growing around them. Ayurveda, the ancient system of Indian medicine dating to at least 6,000 BCE, recommends a profusion of nature-based medicines for every health problem known at the time. Ötzi the Iceman, Europe's oldest mummy, carried with him on his final journey assorted agrarian medicines, including two medicinal mushrooms.[30] In 450 BCE, Hippocrates recommended one of these same species found with Ötzi as an anti-bacterial and cauterizing agent.[31] Traditional Chinese medicine, after 3,500 years of practice, prescribes around 13,000 natural remedies, the bulk of which are plant- and mushroom-based.[32]

These traditional medicines were discovered in many ways; sometimes by trial and error, sometimes by observations of other animals, and other times by the reception of esoteric knowledge gained in meditative states. At any cost, these cures were hypothesized, tested, and found to be effective. Hundreds, if not thousands, are still in use today by peoples around the world. Though so much of what humanity has known of natural remedies has been lost and forgotten over the millennia, we are every day rediscovering the therapeutic properties of the flora and fungi around us. It seems that when we look closer at nearly any plant or mushroom we find that a cure is hidden within its filaments.

A CONTINUING TRADITION

Prevalence of Modern Traditional Medicine Use

Though the efficacy of natural treatments is questioned or outright dismissed by many, the traditional use of herbal and fungal medicines is still in wide use today. In Europe, Germany has one of the highest rates of herbal medicine use among the general population, with an estimated 75% of Germans reporting use of plant-based treatments.[33] Almost 12% of Indians claim traditional medicine to be their primary healthcare option,[34] and the Indian government established an official organization in 2000 to research and coordinate the use of herbal remedies.[35] In a nationwide survey conducted in the US in 2015, over a third of the 26,000 respondents claimed to use herbal medicines.[36]

This is but a small sampling of the global population, and indeed there are areas where the use of plant medicines is not as widespread, such as in South Africa, Russia, and Mexico where less than 1% of the population reports traditional medicine use in a twelve-month period.[37]

It is easy to see that, despite the abundance of prescription and over-the-counter medications available today, many people still seek and utilize the curatives provided by nature. As our knowledge of human physiology and floral biochemistry has grown, so has our knowledge of what is effective and ineffective in terms of natural medicine.

The Folk Roots of Modern Medicine

Modern pharmacology began by and continues in finding inspiration through natural medicines. Several of our present-day prescription and over-the-counter medications originated from plant and fungi sources. Among the most well known is the antibiotic penicillin, derived from the food-spoiling *Penicillium* fungi. Poppies, yielding the popular drug opium, led to the discovery of our most powerful modern painkillers, including morphine, codeine, and methadone. Aspirin has its roots in the traditional use of various tree species, such as white willow and birch, which contain the anti-inflammatory salicylic acid.

These are but a sampling of the many medications either inspired by or simply extracted from flora that has been used in traditional medicine for hundreds and perhaps thousands of years. In a 1988 publication, pharmacognosist Norman R. Farnsworth described 119 plant-based medications, stating that 25% of all U.S. prescriptions for the twenty-five years prior were extracted from plants.[38] He goes on to explain that scientists have studied the chemical constituents of 35,000 plant species. This may seem like a great many, but his argument was that there are an estimated 250,000 to 750,000 plant species in the world, rendering us mostly ignorant of the beneficial compounds these organisms may contain.

As it is, each year more research is done in the realm of pharmacology, and we are continually learning of the curative properties of novel plant and fungi species.

SOUL SICKNESS: HOW MUSHROOMS ARE BEING USED TO HEAL OUR MINDS

Nature-derived medications are not only used to treat bodily ailments but those of the mind as well. *Piper methysticum,* a species of pepper plant from the Pacific Islands used traditionally throughout Polynesia, is showing promising potential in the treatment of anxiety disorders.[39] Saint John's wort has been used for thousands of years, prescribed by Ancient Greek physicians for multitudinous ailments, including driving away evil spirits. Today, this same *Hypericum perforatum* is being researched to great effect in the treatment of mild to moderate depression.[40]

In recent years, the pervasiveness of mental illness has become markedly more conspicuous. The consequences of our collective discontent are manifesting not only in ourselves, but in the planet we live on as well. Forests are being ravaged by industrialism, biodiversity is rapidly decreasing, the air we breathe is thickening with pollutants, the oceans are acidifying, and the list goes on and on. During a 1992 lecture at the Earth Trust Benefit in Los Angeles, California, Terence McKenna gave a succinct description of our present conundrum:

"Many people are in anticipation of a kind of apocalypse- a kind of complete breakdown of social institutions and ideals- and I must say to you the apocalypse is not something which is coming. The apocalypse has arrived in major portions of the planet, and it's only because we live within a bubble of incredible privilege and social insulation that we still have the luxury of anticipating the apocalypse."[41]

However, as our cumulative delirium continues to grow, the stigmatization of psychiatric disorders is falling away, and compassionate outlooks and treatments are starting to become the societal norm.

One of the most propitious of these developing treatments is the use of psychotropic drugs. LSD, MDMA, ayahuasca, cannabis, and psilocybin are all seeing a revival of research in recent years. They are being tested with heartening results in the treatment of a slew of psychiatric illnesses, including addiction, depression, PTSD, OCD, and anxiety. What once was blindly considered taboo, illegal, immoral, and even demonic is now giving way to untainted outlooks of benevolent long-sightedness.

McKenna's advice on avoiding the apocalypse that is knocking at our door was to reconnect with nature, especially through the use of psychedelics. As we look toward the future of psychotropic research, it seems that heaven may arrive sooner than the hell many of us have been expecting.

THE PURPOSE OF THIS BOOK

The authors have undertaken the writing of this book for a few main reasons. When we took a thorough inventory of our motivations and asked, "Truly, why are you doing this?", the loudest answer, overpowering all others, is that we love magic mushrooms. They have influenced our lives in a profound way, and we owe much of our current understanding of the world to the handful of experiences we have had with them.

Like Terence McKenna, we are of the mind that psychedelics can save our world and heal our societies. The expansive effect they have on consciousness, allowing us to see outside ourselves, outside our communities, even outside our dimension, might be that which can draw us back from the edge of oblivion and guide us back to our Edenic homeland.

EDUCATION

The authors hold knowledge to be one of the greatest assets in the uphill battle against the inquisition of our collective insanity. Knowledge on any matter adds fuel to the fire of our consciousness and allows us to fight against the darkness of societal ignorance. In writing this book, we hope to do our own small part in dispelling some of the misinformation that has accumulated around the use of psychotropics, consequential of their undeserved criminalization. We will steadfastly avoid adding anything not based in fact to the pool of rumors surrounding the use of magic mushrooms, expressly stating when a matter is of our own trivial opinions. As many trusted sources as we could find will be listed in the terminal references section so that you may further research the topics discussed and draw your own conclusions about each issue we will cover.

DISCLAIMER
Please note: psychotropic plants and other hallucinogenic substances may be harmful to health, and in many jurisdictions throughout the world, it is illegal to possess and use such substances.

Readers are advised that they use such substances entirely at their own risk. The authors and the publisher of this book disclaim liability for any adverse effects resulting from the use or possession of any psychotropic plant or other hallucinogenic substance that is discussed herein.

ADVOCACY OF SHIFTING PARADIGMS

We authors bounce like ping pong balls between abject despair and magnificent optimism when considering our global situation. On one hand, it seems that every direction we look there is the desecration and destruction of nature, the deterioration of the human spirit, and the disregard for all forms of life. Then, on the other end, we see our fellow humans striving to lift ourselves by means both psychic and somatic; spiritual and technological; ethereal and material. We hear a chorus of inspired crusaders calling for positive change rising above the clamorous cries of confused suffering issuing from those who would watch our planet rot.

Bob Dylan's tremorous proclamation that the times are a-changin' is an old truth, but one that holds special importance now when those changes can be either disastrous or miraculous. Our wish is that this book may affect some small benefit to our global society; that it may serve to embolden those curious adventurers seeking to explore other realms of consciousness and to assuage the fears of those whose thinking on magic mushrooms has been guided by the authoritarians' insistence of danger and degradation.

GRATITUDE TO MUSHROOMS

Although the authors are not regular trippers, and though we do not actively seek out psychedelic experiences, we know that whenever the time is right a good mushroom journey is usually the perfect kick to our egos and boost to our souls that we needed. We believe psychotropic mushrooms to be very benign, beneficial, and largely misunderstood organisms who deserve a closer look by the larger populace and especially by the governmental bodies who forbid their use and cultivation.

Our experiences with magic mushrooms opened our lives to a way of thinking we had never imagined before and instilled within us a deep and lasting sense of peace. Though we seriously doubt they expect much or anything in return for their wisdom, we offer this book as thanks for the betterment these little fungal mages have bestowed upon our lives.

ANCIENT THEORIES

The following section contains many hypotheticals and is included simply to give you, the reader, a more thorough awareness of the schools of thought that have grown up around magic mushrooms. Our understanding of pre- and ancient history is tenuous at best, based much more on suppositions than hard facts. As such, we have no hard evidence that humans of antiquity participated in any ritual or recreational mushroom use, but the theories are nonetheless interesting for those familiarizing themselves with psychotropic fungi. The authors have attempted to order the theories that follow in as much a hypothetical chronology as is possible, and will later advance to recorded historical use.

THE STONED APE THEORY

Terence McKenna is a big name in the world of psychedelic exploration but less so in the realm of accepted prehistorical theories. In his 1992 book *Food of the Gods,* McKenna presented what has become known as the "Stoned Ape

Hypothesis."[42] According to this theory of human evolution, the changes that occurred in the brains of *Homo erectus* that resulted in the evolutionary branching off of the unique *Homo sapiens* species were precipitated by the addition of psilocybin mushrooms to our ancestors' diets.

McKenna argues that human consciousness evolved 100,000 years ago during a desertification phase of the African continent. As the environment changed, ancient humans were forced to find other food sources and serendipitously stumbled upon psilocybin mushrooms growing in the African plains. The inclusion of magic mushrooms in the *Homo erectus* diet drove several psychological and physiological transformations. According to McKenna, mushrooms increased the species visual acuity, leading to more successful hunting tactics. Consequently, this larger diet, along with psilocybin's potential to enhance the sex drive, led to greater reproductive success. Ultimately, the Stoned Ape theory suggests that the ingestion of mushrooms by ancient humans climaxed with the development of language, culture, and religion, elevating our species from its bestial origins and into the new dimension of civilized humanity.

Due to a lack of evidence and a smattering of contrary observations, Terence McKenna's theory on human evolution has either been largely ignored by the scientific community or received heavy criticism. Indeed, there is no proof that our ancient ancestors had any interaction with psilocybin mushrooms prior to the last couple centuries, but it is a fascinating hypothesis to consider.

The first physical evidence we have that lends credence to theories regarding ancient mushroom use is in the form of murals painted on rocks and cave walls. Their ages range from 3,500 to 11,000 years, and though it is a matter of debate what is actually represented in these images, the inference that they depict psychotropic mushroom species is as plausible as all other theories.

Tin-Tazarift, Tassili, Algeria

The engravings and paintings of Tassili, Algeria contain several fungoid objects among the 15,000 different pieces of prehistoric cave art. Dispersed amidst the conglomeration of 5,000 to 11,000-year-old paintings are various scenes hypothesized as showing anthropomorphic fungi and mushroom shamans. The most striking of these is a figure at Matalem-Amazar, which appears to be a masked figure with mushrooms sprouting from the perimeter of its body.

Giorgio Samorini was the first to give an in-depth examination of the Tassili art in terms of psychotropic shamanic practices, speculating that many of the forms depict *Psilocybe* and *Amanita* species, as well as trance states and shamanic rituals.[43]

Selva Pascuala, Cuenca, Spain

Throughout the Iberian region of Spain are thousands of instances of rock and cave art. In Selva Pascuala, there is a fascinating 5,000-year-old piece featuring an obvious bull accompanied by a row of thirteen mushroom-like objects. We can't say with certainty that these are indeed mushrooms, but they sure do look like them, with slender stems supporting bulbous caps. This scene has been theorized as portraying the species *Psilocybe hispanica*,[44] known to grow in the region in which the art is located and frequently found in animal dung.

Pegtymel', Chukotka, Russia

Along the banks of the Pgtymel' River in Chukotka, Russia, archaeologists uncovered a series of rock carvings dating to around 1,450 BCE.[45] Here are clear representations of hunts, of reindeer, and of boats, along with forms which appear to be anthropomorphized mushrooms or else humans carrying large mushrooms. These mushroom-like illustrations, in place among the vitally important hunting and sailing scenes, suggest that the peoples of this region were engaged in the ritual use of *Amanita muscaria*. Similar scenes occur throughout Scandinavia, as featured in rock art of Norway and Sweden.[46]

SOMA: THE DRINK OF IMMORTALITY

In Vedic literature, a mysterious floral entity and its preparation are referred to as soma. Ralph T.H. Griffith translated a verse invoking its use from the 2,500-year-old Rigveda scripture:

We have drunk soma and become immortal; we have attained the light, the Gods discovered.

Now what may foeman's malice do to harm us? What, O Immortal, mortal man's deception?[47]

Many theories have been suggested as possible identities of the source of soma, including cannabis, opium poppy, or simply honey. In 1968, R. Gordon Wasson and Wendy Doniger postulated in their book *Soma: Divine Mushroom of Immortality* that soma may have been *Amanita muscaria*.[48] However, the effects of fly agaric do not correlate well to those attributed to soma in ancient literature, and the theory has been widely criticized. Terence McKenna, countering Wasson's hypothesis, proposed *Psilocybe cubensis* as the magical substance known as soma, suggesting a link between the coprophilic psilocybin species and the sacred view of cows held throughout much of ancient Indian scripture.[49]

Neither of these mushroom theories has met with much support, and it is currently held that the most likely candidates for the original soma were opium poppy, cannabis, and ephedra.[50]

TEONANÁCATL: MUSHROOMS IN THE MAYAN AND AZTEC CULTURES

The most credible records of an ancient use of magic mushrooms are found in the history of the Mayan and

Aztec peoples of Mexico and Central America. Their descendants, being the only known modern groups to have a traditional practice of psilocybin ritual, give credence and congruency to the suggested link between ancient Mesoamerican art and sacred mushroom use.

Sculptures of mushrooms with animalistic and anthropomorphic features found in Mesoamerica dating to 3,000 BCE suggest a ritual use of psilocybin mushrooms that is several millennia old. Several murals throughout the region also depict mushrooms in ritualistic and religious settings, as in the Tepantitla mural in Teotihuacán, which displays a god surrounded by priests presenting offerings of fungi. The Dresden and Madrid codices of the Mayans depict mushrooms in various scenarios and date to around 1,000 CE.[51]

These instances are not conclusive proof that the Americans of antiquity did indeed use psilocybin species in any of their traditions, but there is no doubt they ascribed at least some degree of importance to members of the fungi kingdom.

EARLIEST
RECORDED USE

THE TROUBLE OF RECORDS

There are a few issues that arise when considering the documented use of magic mushrooms. Most apparent is that records are quite often simply lost to time. This can happen through an assortment of ways. Testimony may just become a cultural obscurity as the years march on, with subsequent generations failing to detail every aspect of their cultures' past. Likewise, natural disasters such as fire and flooding are apt to destroy the more ephemeral instances of the written word and have done so many times since we began our chronicles of civilization. More nefariously, the invasion of nations by outside conquerors often tragically results in a purposeful destruction of cultural records. Such was the case of the Library of Alexandria, the Library of Antioch, and the bulk of the Aztec and Mayan records, which were destroyed by Catholic priests seeking to subdue the indigenous religion they considered satanic.

On the other hand, cultural history may never be recorded in the first place. Our lack of information regarding prehistoric mushroom use is probably due less to the intentional or accidental destruction of heritage, but owed more to the fact that these peoples lacked a writing system. This is the case with many of the types of cultures thought to practice entheogen use, as in the Koryaks and other Siberian tribes known to practice *Amanita* shamanism. Furthermore, even if these cultures do have a system for recording their cultural practices, their religious practices may be excluded to protect their esoteric knowledge.

Whatever the case may be, we have only so much on which to base our theories of historical psilocybin use.

THE SURPRISINGLY BRIEF HISTORY OF RECORDED USE

We can't say definitively when in humanity's history magic mushroom use began, and our oldest documentations date back to a relatively recent four centuries ago. The theories pertaining to more ancient utilization of psychedelic fungus are interesting pieces of food for thought, but in the end are merely hypotheses. What we know for sure is attributable to a few honest chroniclers who, as best as we can tell, faithfully reported their sightings of magic mushroom practices.

Annals of the Aztecs

Upon the arrival of Spaniards to Mesoamerica, Western chroniclers, mostly clergymen, began recording magic mushroom use present among the Aztecs. The earliest recorded mention of psilocybin use is found in Bernardino de Sahagún in his comprehensive work, *The Florentine Codex: General History of the Things of New Spain:*

"...they possessed a great knowledge of plants and roots, and they were acquainted with properties and virtues of them; these same people were the first to discover and use the root which they called *peiotl*, and those who are accustomed to eat and drink them used them in place of wine; and they did the same with those which they call *nanacatl*, which are harmful mushrooms which intoxicate in the same way as wine..."[52]

Others wrote of mushroom use as well but often cast it in a negative light, as the Catholic frays who reported on much of Mesoamerican culture regarded the natives' religion as blasphemous. Consequently, as the Spanish conquered the peoples of Mesoamerica, they destroyed many of the natives' historical records. It will always be unknowable what was lost in those chronicles, so we can only guess as to how far back ritual psilocybin use actually went among the indigenous Mesoamericans. Nowhere else in the world has there been found recorded traditional use of psilocybin species.

Siberian Use of *Amanita muscaria*

Aside from what we can infer about European *Amanita* use from the assortment of prehistoric rock carvings, our official accounts of their ingestion begins in the 17th century. In 1657, a Polish prisoner of war described the Ket people's use of "fungi in the shape of fly-agarics" to produce a state of drunkenness.[53] Following this initial report, there have been various explorers and ethnographers to document *A. muscaria* use by various Siberian groups, sometimes in shamanic settings and otherwise recreationally by laypeople. An account by the Swedish Filip Johann von Strahlenberg from his 1730 book *An Historico-Geographical Description of the North and Eastern Parts of Europe and Asia* contains a description of both the ingestion of *A. muscaria* decoctions and the drinking of urine of those who had recently become intoxicated by the mushroom:

"The Russians who trade with them [Koryak - a tribe on the Kamchatka peninsula], carry thither a Kind of Mushrooms, called in the Russian Tongue, Muchumor, which they exchange for Squirils, Fox, Hermin, Sable, and other Furs: Those who are rich among them, lay up large Provisions of these Mushrooms, for the Winter. When they make feast, they pour water upon some of these Mushrooms and boil them. They then drink the Liquor, which intoxicates them; The poorer Sort who cannot afford to lay in a Store of these Mushrooms, post themselves on these occasions, round the huts of the rich and watch the opportunity of the guests coming down to make water. And then hold a wooden bowl to receive the urine which they drink off greedily, as having still some virtue of the mushroom in it and by this way they also get drunk."[54]

Maybe the most popularized usage of fly agaric involves the supposed habit of Vikings in their preparation for combat. However, the popular account that the Norse people used this mushroom to enter their berserker state of battle frenzy is unfounded speculation, first recorded by Samuel Ödman in 1784. Ödman's postulations are wholly unsubstantiated theories, with little to no basis in historical fact.[55]

In North America, documented traditional fly agaric use was first written in 1978 by the Anishinaabe elder Keeway-dinoquay Pakawakuk Peschel in her book *Puhpohwee for the People: A Narrative Account of Some Used of Gungi Among the Ahnishinaubeg.*[56] Further North American use by natives of British Columbia was described by Richard Evans Schultes, Albert Hofman, and Christian Rätsch in their collaborative 1979 work *Plants of the Gods: Their Sacred, Healing, and Hallucinogenic Powers.*[57]

Contemporary use of *Amanita muscaria* is purported among Lithuanians [p.43-44] in remote areas of the country, who use them in some wedding ceremonies.[58] Recreational and ritual usage is not widespread in modern times, but with the distinct outlawing of psilocybin in the U.K., fly agaric experiences began to see a rise in popularity.[59]

Western Use

Although the Spanish explorers had documented the use of magic mushrooms among the Mesoamerican natives, they didn't become popularly known to Western society until the 1950s. This lack of notoriety was probably deeply rooted in the mushroom's categorization as an evil agent of magic by the Catholic clergy who first wrote about them. The Spanish exploration of South and Central America was largely an economic and religious conquest, during which resources were stolen, indigenous populations were decimated, and those that survived were subject to the ultimatum of converting to Catholicism or being subject to harsh punishment. As the native's texts, customs, and cultures were destroyed, what was left of their religious practice was forced to continue in practice only clandestinely. Ceremonies which had been grandiose and public were relegated to an underground secrecy, sometimes being forgotten altogether.

Surely, any mention the conquerors gave of magic mushrooms amongst their countrymen was cast in a light of bigotry and immorality. The "heathen practices" would be ridiculed and vilified, assigning a gross taboo to the native's ancient spirituality. Ultimately, talk of magic mushrooms grew quiet, with only faint whispers of their existence and use circulating over the next several centuries. There is a recorded case of an accidental experience with a psilocybin-containing species from London in 1799,[60] but this led to no further serious investigation of their neurotropic properties.

A further early western account of magic mushroom experimentation is recorded by A.E. Verrill in *Science* magazine during September 1914.[61] In this article, titled "A Recent Case of Mushroom Intoxication," Verrill recounts the descriptions of one Mr. W, a botanist in Oxford County, Maine, and his niece in their experimentation with a *Panaeolus* species, listed in the article as *P. papilionaceus*, however, this species does not contain any psychoactive compounds. Since the pair reported all the effects now known to be common to psilocybin intoxication, it is unlikely that their experience was a mere placebo response to a non-hallucinogenic species, with more probability resting on the assumption that this was a case of misidentification.

Though earlier documentations were made in prior years,[62] it wasn't until 1940, with Richard Evans Schultes' publication of "Teonancatl: The Narcotic Mushroom of the Aztecs",[63] that Westerners started to open their eyes to the use of psilocybin fungi. Schultes' research caught the interest of banker and ethnomycologist R. Gordon Wasson. Together with his wife, Valentina Pavlovna Guercken, Wasson began traveling to Mexico in 1953 in search of the fabled sacred mushroom. In 1955, the Wassons participated in a healing vigil known as a *velada* in Oaxaca, Mexico, under the guidance of Maria Sabina. Thus, the Wassons became the first Westerners known to have experienced a magic mushroom trip.[64]

Gordon Wasson published multiple accounts of his experience, and his article "Seeking the Magic Mushroom" in *Life* magazine kickstarted a huge Western interest in the psychotropic fungi. Against her wishes, curandera Maria Sabina's name and location were published in certain works by Wasson, and her town of Huautla de Jimenez was soon flooded by psychedelic tourists. Perhaps for this reason, perhaps for others, Sabina's house was burnt to the ground, and much of her life after Wasson's publications was visited by suffering.[65] This is quite tragic, and Wasson was never known to regret his disclosure of the medicine woman's private details, but history is very often a record of tragedies and their subsequent tidal wave of effects.

In this case, that wave was an exponentially increasing interest in not only psilocybin mushrooms, but psychotropic drugs in general. Psilocybin and psilocin were identified and extracted as the active components of magic mushrooms by Albert Hoffman in 1958, and in 1960, Timothy Leary and Richard Alpert began work in the Harvard Psilocybin Project. Around this same time, the CIA began their own experiments into mind control using psilocybin.[66] These controlled dosings fed the flames of interest among young adults, often the subjects of these experiments, and psilocybin use soon became ingrained into the hippie counterculture movement of the 1960s.

Urged by advocates such as Timothy Leary to "turn on, tune in, and drop out," many people began personally experimenting with spiritual and recreational use of psilocybin. Interest in psychotropic drug use continued to grow until the implementation of the Controlled Substances Act in 1971. This legislation effectively banned psilocybin use in any measure, labeling it a drug with a high potential for abuse, no accepted medical use, and no safe applications, along with nearly every other psychotropic drug currently known at the time.[67]

Although a decade of studies had shown beneficial and promising results using psychedelic substances as psychotherapeutic drugs, research in the U.S. halted in 1971 and had ceased worldwide by the 1980s. This, of course, would not eradicate individual use of these drugs, and psilocybin mushrooms continued to be used for recreational, spiritual, and therapeutic purposes up to the present day.

Current Use and Modern Research Endeavors

In spite of the efforts of authorities to end it, the counterculture movement of the 1960s did not die; rather, it retreated quietly to just beneath the surface, slowly spreading like the mycelial threads of our protagonistic fungi. Many hippies continued to hold to their values of connectivity with nature, love of each other and of the planet, and psychedelic drug use. Parts of the movement branched off, mutated, and evolved into rave culture, wherein the use of psychotropics continued to maintain a special interest in a drug-induced expansion of consciousness.

These old hippies and these new ravers, as well as physicians, psychiatrists, business leaders, and general activists, have continued advocating for the decriminalization of psychedelics. The first big win in this area was the 1996 legalization of medical cannabis in California. Cannabis reform gradually snowballed for the next two decades, as more states and some countries have either decriminalized or fully legalized marijuana use.

This changing perspective on cannabis use has strengthened the arguments of psilocybin advocates, and new research began to be conducted on the psychotherapeutic benefits of psilocybin in the late 1990s. In 2006, Dr. Francisco A. Moreno of Tucson, Arizona led the first FDA-approved study of psilocybin since its ban in 1970.[68] Since then, several more studies have gained the approval of the FDA, and twice, in 2018 and 2019, psilocybin has been designated by the FDA as a breakthrough therapy for depression.[69,70] In 2019, Denver, Colorado became the first US city to decriminalize psilocybin, followed shortly thereafter by the California cities of Oakland and Santa Cruz.[71] Currently, three states - California, Iowa, and Oregon - are considering ballot measures or legislation that would decriminalize mushrooms and allow for further medical research.

FIRST RESEARCH (MID-1900S)

A BANKER, A SPY, AND A BREAKTHROUGH

R. Gordon Wasson is a surprising figure in the world of psychonauts. After all, who would expect a vice president of a major American bank to extol the virtues of hallucinogens? However, that's exactly what happened. Wasson first became interested in mycology during his 1927 honeymoon to the Catskills Mountains with his wife, Russian-born Valentina Pavlovna Guercken. While at their rural New York retreat, Valentina found edible mushrooms she knew from her childhood in Russia. When Gordon refused to eat them, a discussion ensued in which the couple supposed there were two types of people in the world: mycophobes and mycophiles. This difference in cultural attitudes toward mushrooms sparked Wasson's interest in fungi, leading him and his wife to coin the term *ethnomycology*.

After reading an article by Richard Evans Schultes offering an account of traditional mushroom rituals by indigenous Mexican peoples, Wasson and his wife began undertaking journeys to Mexico to study the Mazatec's various fungal beliefs. In 1955, while traveling with photographer Allan Richardson, Wasson met the medicine woman Maria Sabina. The two men took part in a healing vigil, during which they consumed psilocybin mushrooms for the first time. Enchanted by the ceremony and the effects of the mushrooms, Wasson met with his wife and daughter the next day, and the two women experienced a psilocybin trip as well.[72]

Wasson returned to Sabina's village for additional veladas over the next few years. In 1956, Wasson was accompanied by a chemist by the name of John Moore. The CIA had begun the MKUltra project in 1953 in which they had been studying the mind-control potential of psychotropic drugs. Having somehow learned of Wasson's "discovery" of the mushroom known as *God's flesh*, the CIA tasked Moore with accompanying Wasson on a mushroom expedition in an effort to isolate the chemical compound responsible for the hallucinogenic effects. Wasson was unaware of Moore's CIA involvement and happily accepted him as a companion, in addition to the $2,000 grant Moore was able to procure. The grant was financed by the Geschickter Fund, a shell run by the CIA.

Mycologist Roger Heim also accompanied Wasson on this expedition, and another mushroom velada was commenced under the direction of Maria Sabina. Upon their return to the U.S., Moore brought with him a bag of *Psilocybe mexicana* mushrooms and began his research into isolating the active components under the direction of the CIA. Roger Heim, meanwhile, had succeeded in growing the same species in his lab using spore prints he had taken during the trip to Mexico. Soon thereafter, Heim sent samples to LSD-inventor Albert Hoffman. Hoffman, along with a team of researchers at the Swiss company Sandoz Pharmaceuticals, successfully identified, isolated, and extracted the psychoactive compounds of the *Psilocybe mushroom* in 1959.[73]

Sandoz began selling pure psilocybin under the brand name Indocybin for use in psychotherapeutic medicine. This, in turn, led the CIA to give up their own efforts to synthesize the mushroom compounds, instead opting to obtain psilocybin through Sandoz; thus following the same route they had taken in procuring the LSD used in MKUltra experiments during the preceding few years.

In the following years, R. Gordon Wasson would continue to study the cultural significance of mushrooms and expand the field of ethnomycology. His research led him to spend the bulk of the years 1963 through 1966 in field investigations throughout New Zealand, New Guinea, Japan, China, India, Korea, Iran, Afghanistan, Thailand, and Nepal. During this time he studied the entheogenic practices of many different cultures while collecting data for his theory on the identity of the *soma* of ancient Vedic literature. His wife, Valentina Guercken, died of cancer in 1958, but not before having been one of the first to shrewdly suggest that psilocybin might have great potential in the treatment of psychiatric diseases and addiction.[74]

The same year as psilocybin's Sandoz synthesis, psychologist Timothy Leary became a lecturer at Harvard University. Soon after, Leary, through a friend, learned of psilocybin mushrooms and in 1960 traveled to Mexico to trip on the sacred shrooms. He then returned to Harvard, and together with Richard Alpert, Aldous Huxley, David McClelland, Frank Barron, and Ralph Metzner, started the Harvard Psilocybin Project. The project was formed as a subset of the Harvard Center for Research in Personality. Leary and Alpert would lead the research, beginning by ordering psilocybin from Sandoz Pharmaceuticals.

The Harvard Psilocybin Project is often described as "a loose set of experiments," which was basically the case. Leary and Alpert began by administering psilocybin to 38 volunteers in comfortable, informal settings. The results were positive from the start. 75% of participants stated that the psilocybin experience was very pleasant, while 69% reported that the trip had caused a significant broadening of awareness. Over 400 people participated in the Harvard Psilocybin Project in its operational years between 1960 and 1962, with 66% of subjects claiming the experience changed their lives for the better.[75]

The Concord Prison Experiment

There were two notable experiments conducted during the two year run of the Harvard Psilocybin Project. First, was the Concord Prison Experiment.[76] The experiment was designed to test whether or not psilocybin-assisted psychotherapy would have a positive effect on reducing the recidivism rates among recently released prisoners. Unlike the CIA's unethical MKUltra psilocybin experiments which forcibly dosed unwitting prisoners, all 32 of the Concord Prison Experiments subjects were volunteers. In this study, inmates of the Massachusetts Correctional Institute undertook a six-week psychotherapy treatment which included doses of psilocybin ranging from twenty to seventy milligrams, counseling sessions, and post-parole planning meetings. After the inmates' release, recidivism rates were measured for the next two and a half years.

Leary reported a drop in recidivism rates of up to 20% and concluded that the study had been overall successful. However, a reevaluation of the experiment's results conducted in the 1990s found that Leary had significantly skewed and misrepresented the stats to shine a favorable light on the benefits of psilocybin therapy. The follow-up study determined that psychedelic-involved psychotherapy is not enough in itself to have a significant effect on personality changes related to the antisocial behaviors that lead to recidivism.

The Marsh Chapel Experiment

The Marsh Chapel Experiment, also called the Good Friday Experiment, was conducted by graduate student Walter N. Pahnke of the Harvard Divinity School.[77] Pahnke designed and conducted the experiment under the supervision of the Harvard Psilocybin Project to test the ability of psilocybin to cause so-called "mystical" experiences. In the study, twenty test subjects were divided into a control and an experimental group. The participants, all undergraduate theology students, were taken to the basement of the Marsh Chapel and administered either psilocybin or niacin pills. A sermon was played over speakers and accompanied by choral and organ music.

Results were collected using a post-study questionnaire and measured factors such as feelings of unity, a transcendence of space and time, deeply felt positive mood, a sense of sacredness, insight and illumination, and persisting positive changes in attitude and behavior. The conclusion was that all participants in the experimental group, those who had received psilocybin, reported definite mystical experiences. Their results ranked high in the categories of unity, transcendence of time and space, paradoxicality, feelings of sacredness and love, and persisting positive changes in behavior and attitude toward self and life. Follow up experiments and similarly modeled experiments conducted since the Marsh Chapel study have largely confirmed and supported Pahnke's conclusions.[78]

CONTROVERSY AND CLOSURE OF THE HARVARD PSILO-CYBIN PROJECT

As word circulated around the Harvard campus of Leary and Alpert's intriguing Psilocybin Project, controversy began to grow amidst the faculty and administrators.

According to the terms of a 1961 agreement between the university and the Harvard Psilocybin Project, only graduate students would be permitted as experiment participants. In breach of this agreement, Leary oversaw the administration of psilocybin to many undergraduates. Additionally, some Harvard administrators found reproach in the fact that the conductors of the experiments frequently took doses of psilocybin during the course of the studies. To this point, Leary argued that it was helpful for the experimenters to be in a similar state of mind as the experimentees as to better understand their experiences.

Despite these deviations from agreed-upon protocols, the Harvard Psilocybin Project continued to run mostly without challenge for the two years following its inception. Then, in February 1962 an article was published in Harvard's student newspaper, *The Harvard Crimson,* calling into question the safety and forthrightness of Leary's project. Leary wrote a letter of reply to the paper in which he advocated the safety and great potential of the psilocybin experiments. This, in turn, resulted from a letter to *The Crimson* from the director of Harvard University Health Services, Dr. Dana L. Farnsworth, who warned of the dangers of psychotropic drugs, mescaline in particular.

Farnsworth's letter led to more outspoken criticism of the Harvard Psilocybin Project, culminating in a private meeting of the members of the Center for Research in Personality on March 14, 1962. During the meeting, many expressed displeasure in the methodology of Leary and Alpert's experiments, claiming that the researchers were not giving enough thought to the possible long-term psychological damage that may come with psychotropic drug use. Leary and Alpert defended their positions and assured the other members that emergency medical services were constantly on standby if at any time they were required.

Unknown to the Center for Research in Personality members, a reporter from *The Harvard Crimson* was in attendance at the meeting and would go on to publish an unflattering account of the proceedings. Soon after, local Boston newspapers began reporting on the controversy, followed by an official announcement that the Massachusetts food and drug division would begin an investigation into the psilocybin project. Through this time, not all Harvard administrators were so critical of Leary and Alpert's work. The university president stated that Harvard would launch no investigation of its own, and other faculty members expressed their unwillingness to inhibit academic freedom by opposing the research.

At the conclusion of the state's investigations into the Harvard Psilocybin Project, the official account dictated that its research could continue so long as a medical professional was present during the administration of the psilocybin. This did not please the dissenters, who continued to fight against the project. Faculty and administrators were concerned about the rising popularity of psychedelic drugs among Harvard students. Mescaline and LSD were beginning to circulate in larger and larger numbers, being used both in unsanctioned psychotherapeutic experiments and for recreational purposes. The dramatic atmosphere surrounding their research prompted Alpert and Leary to take a sojourn to Mexico, at which point they started their Zihuatanejo Project. Upon returning to Harvard in 1963, both were fired shortly after; Alpert's termination based on accusations that he had provided psilocybin to undergraduates and Leary's on the grounds of having been absent from his lecturing duties. Thus, without its leading researchers, the Harvard Psilocybin Project was officially disbanded.[79]

THE EFFECT OF THE WAR ON DRUGS

Of the aftermath of Harvard's psilocybin studies, perhaps the most far-reaching consequence was the rise in popularity of psychotropic drugs. There were hundreds of participants in the Harvard Psilocybin Project, through which rumors of the profound effects of these substances reverberated far and near. Moreover, Timothy Leary and Richard Alpert were far from done with their psychedelic advocacy, gaining a much larger following among the hippie community over the next decade. In 1962, the two psychonauts founded the International Federation for Internal Freedom with its headquarters in Zihuatanejo, Mexico.[80] At this time, they turned their attention more toward the spiritual use of LSD, and over the course of the 1960s would become two of its most powerful advocates.

Leary, now more psychedelic guru than psychiatric researcher, amassed a huge following during these years. As psychedelic drug use rose in popularity, the U.S. government began to limit its legality. Through various measures in the '60s, the production of LSD became first restricted, then outlawed. Its use was strictly controlled and limited only to approved medical applications.

Psilocybin was first regulated under the Drug Abuse Control Amendments of 1965, issued to control the production, sale, and possession of hallucinogens, sedatives, and stimulants. In 1967, the U.S. Food and Drug Administration with the National Institute of Mental Health formed the Psychotomimetic Advisory Committee for the inspection and processing of all requests to use psychoactive drugs.[81] With this, the crackdown on psychedelics then began in earnest, fueled in part by the CIA's desire to control the movement and use of psychoactive substances, coming into full force with the 1971 *Convention on Psychotropic Substances* United Nations treaty.[82] This treaty, still in effect today and including 183 member states, banned the use of psilocybin worldwide.

CEASE OF RESEARCH

Under the *Convention of Psychotropic Substances*, most psychedelics fall under a Schedule I classification, deemed to have no acknowledged therapeutic value and creating a serious risk to public health. Although the treaty does not outright ban any natural psychotropics, regional governments have based laws criminalizing these agents based upon the Convention's framework. Mushrooms are included in this ban by being considered containers for or preparations of psilocybin.

With this illegalization came a ban on virtually all research in any capacity, with the last FDA-approved psilocybin study conducted in 1977.[83] So, although psychotropics such as LSD and psilocybin had shown promising psychotherapeutic potential, studies were abandoned, and researchers were forced to turn over their supplies of these drugs to governmental authorities. This ban on research has persisted in most member states through the bulk of the last fifty years.

A SUBTLE REAWAKENING

Despite the cultural and legal taboos placed on magic mushrooms since their criminalization in the 1960s and '70s, we are starting to see a resurgence of interest in their medical applications. For those who believe psilocybin to be a truly beautiful and blessed thing, this is a very welcome change of pace. In the authors' opinions, to outlaw any naturally occurring organism for personal use is in itself a criminal act, and we are nothing but heartened that we are seeing a shift in perspective regarding these psychotropic fungi. Not only are the medicinal potentials of magic mushrooms extremely valuable in the realm of mental health, but each individual's own mind should be theirs alone to do with as they please. To once again quote Terence McKenna:

> "People's minds, like their bodies, must be a domain free from government control."[84]

WHAT CAUSED THE MUSHROOM REVIVAL?

Though progress has been slow, advocates of psychedelics haven't stopped fighting for the last fifty years. The decade of the 1970s saw continued support for hallucinogenic mushrooms and was the era of the first-ever International Conference on Psychotropic Fungi,[85] held at Millersylvania State Park in Washington state. This conference was co-founded by Jonathan Ott, Preston Wheaton, and Tim Girvin. Speakers and presenters included the father of ethnomycology R. Gordon Wasson, biochemist Jeremy Bigwood, anthropologist and mushroom historian Gastón Guzmán, mycologist Steven Pollock, and modern-day mushroom guru Paul Stamets. At this conference, researchers presented new findings regarding psychotherapeutic uses, physiological discoveries, and cultivation techniques for growing psilocybin species.

It was followed in 1977 by the Second International Conference on Hallucinogenic Mushrooms, hosted in Fort Worden, Washington and accredited by the Washington State Medical Association.[86] Many of the same presenters were in attendance at this conference. Also presenting were the director of the Harvard Botanical Museum, Richard Evans Schultes, Albert Hoffman, and drug addiction researcher and psychiatrist Norman Zinberg. Approximately 250 people attended this conference, including chemists from the U.S. DEA and members of the Royal Canadian Mounted Police. Selections from this conference were later published in the 1978 book *Teonanácatl:*

Hallucinogenic Mushrooms of North America by Jonathan Ott and Jeremy Bigwood.

Several conferences similar to these would be held over the next few decades and carry on into the present, with topics ranging from the general use of psychedelics to more specific considerations of their entheogenic virtues. The ranks of psilocybin scientists and activists, which started with the likes of Schultes, Wasson, Hoffman, and Guzmán, would continually be supplemented by newcomers to the field. Each of these advocates helped to advance our understanding of psilocybin and its acceptance in the wider culture, leading us now to the present during which more psychedelic drug research is being conducted than at any time in the past.

The authors owe a brief word of thanks and mention to all of these proponents and will give a brief outline of the advocacy of a very few in the following section. There are many more advocates than we can name, and these are but a selection of some of the greatest voices in psilocybin activism.[87]

First Wave Pioneers

These are the researchers who were the first to consider the science and implications of sacred mushroom use, greatly adding to our first understandings of the magical fungi and further inspiring future supporters and analysts.

Richard Evans Schultes

Schultes, born in 1915, was one of the foremost 20th Century researchers in the field of ethnobotany. He was one of the leading voices responsible for identifying the Aztec teonanácatl as a mushroom and was the first researcher to scientifically document the preparation of the Amazonian hallucinogenic *ayahuasca*. In his travels through South America, he identified 300 species of plant unknown to Western science at the time, as well as being an important figure in the study of the botanically-derived muscle relaxant drug *curare*. Schultes was also one of the first to draw attention to the destruction of the Amazonian habitats. Although Schultes dismissed most of the psychonaut community, his continued support of entheogenic plants and lectures and writings pertaining to such carried the movement throughout the century until his retirement in 1985.

Robert Gordon Wasson

We've already touched on Wasson's contributions to the study of magic mushrooms, but his impact is deserving of further mention. As the founder of the field of ethno-mycology and one of the originators of the term *entheogen*, Wasson's study of sacred mushrooms not only brought our collective attention to the subject but inspired many more to add to this ever-expanding area of knowledge. His historical and geographical cataloging of global ritual mushroom use shed light on what was altogether a largely

misunderstood or unknown category of human spirituality. Wasson's theories provoked thought and debate
among academics that continues to evolve to this day.

Gastón Guzmán

The bulk of our knowledge of the many psilocybin-containing mushroom species is owed to Veracruzian
Gastón Guzmán's study. Guzmán was the first to describe
more than half of the species known to contain psilocybin,
as well as their use, distribution, and chemistry.

Rolf Singer

German-born Rolf Singer classified over 2,600 species
of fungi throughout his career and greatly added to our
knowledge of the *Amanita* genus. After Valentina Pavlovna
Guercken, he was the second to suggest that psilocybin
mushrooms may have therapeutic potential, based on his
observations of indigenous Oaxacans. Over his lifetime,
Singer wrote 440 papers regarding the studies of fungal
nomenclature and systematics, cultivation, ecological
value, and entheogenic use.

Roger Heim

Heim, the French mycologist who accompanied R.
Gordon Wasson on some of his trips to Mexico, was the
first to cultivate psilocybin mushrooms and, through his
collaboration with Albert Hoffman, was instrumental in
the identification of the chemicals psilocybin and psilocin.
In addition to his contributions toward biodiversity

preservation, he published several papers on magic mushrooms that increased our understanding of the species.

Second Wave Advocates

Following on the tide of research begun by the first wave of magic mushroom researchers, these dedicated scientists, activists, and psychonauts helped bring psilocybin to the mainstream and enabled us to reach the point of comprehension at which we stand today.

Jonathan Ott

Ott was born just six years before Wasson's first mushroom trip but has gone on to work alongside some of the leading names in the field of ethnomycology, including Wasson himself, Albert Hoffman, and Richard Evans Schultes. He was one of the original co-organizers of the first conferences on psychotropic fungi and has written or co-authored thirteen books on the subject of entheogens. In present times, Ott still conducts yearly seminars on entheogenic studies and applications.

Paul Stamets

Though some in the mycological world decry Paul Stamets as a charlatan and purveyor of unproven remedies, he was an early proponent of mushrooms who helped popularize the use of fungi in general, not only psilocybin species, in Western culture. Stamets has added a wealth of knowledge to mushroom cultivation techniques and is an outspoken advocate of fungi, proclaiming his belief that "mushrooms will save the world."

Steven H. Pollock

Pollock was an enthusiastic mycologist who discovered four previously undescribed psilocybin species and wrote one of the first books on magic mushroom cultivation. One of his most outstanding achievements, and a unique addition to the realm of magic mushrooms, was the isolation and cultivation of a *Psilocybe* strain that grows in an unusual and unrecognizable truffle form. Producing this strain, he was able to sell magic mushrooms "hidden in plain sight" across the world, and in 1979 Pollock's company Hidden Creek became the largest magic mushroom vendor in the world.

Third Wave Advocates and Beyond

Coming later to the mushroom game than the second wave of advocates and persisting to current times, these researchers and promoters have continued the consciousness crusade through their books, publications, and outspoken support of psilocybin mushrooms and other entheogens.

Giorgio Samorini

Samorini was the first to advance the theory that the cave art in Tassili, Algeria signified the ancient use of magic mushrooms. He has studied entheogens worldwide and published several books on the subject, currently working as the editor-in-chief of the scientific publication *Eleusis: The Journal of Psychoactive Plants and Compounds.*

Christian Rätsch

Based out of Hamburg, Germany, Christian Rätsch has studied shamanism and psychoactive plants since the age of ten. His research into indigenous Mexican entheogen use has greatly added to our knowledge of the subject. Among his greatest contributions is the comprehensive *The Encyclopedia of Psychoactive Plants,* which details the history, cultivation, and usage of over 400 psychotropic species of plants and mushrooms.

Earth and Fire Erowid

The founders of Erowid Center have done a great service by hosting an absolute wealth of unbiased drug information for the last twenty-five years. Their website, erowid.org, provides an unskewed informational database pertaining to over 700 psychoactive substances, including user reports, soundly-sourced scientific insights, and an online library containing an abundance of books related to psychedelic research. With a neutral, data-centric approach to psychoactives, Erowid has given the world access to an impartial source of information related to both the positive and negative aspects of drugs and their use.

Dennis and Terence McKenna

The brother power-duo of Dennis and Terence McKenna contributed an amazing amount of philosophy, theory, and knowledge to our understanding of the psychedelic experience and its significance. The two were the first to offer a reliable magic mushroom cultivation technique for home growers with their 1976 book *Psilocybin: Magic Mushroom Grower's Guide*. Terence became a popular lecturer in the 1980s and continued speaking on the benefits and importance of psilocybin and the psychedelic experience until his death in 2000. Authors of several books, the works of the McKenna brothers have led to an increased interest and acceptance of magic mushrooms and entheogens in general. Dennis McKenna continues to speak on the subject and has been featured in several popular documentaries and podcasts focused on psychedelic advocacy.

Ryan Munevar

Among those leading today's movement to restore our right to magic mushroom use is the director of Decriminalize California, Ryan Munevar. With a mission statement proclaiming the promotion of "research and education concerning the healing and consciousness expanding nature" of psilocybin fungi and their return to legal use, Decriminalize California is making great strides in the current push to remove the criminality and stigma associated with magic mushrooms.

Amanda Fielding

Founder of the Beckley Foundation, Amanda Fielding has done tremendous work in lobbying for drug policy reform. She has been called the "Queen of Consciousness," having enabled through funding, directing, and co-authoring the continued study of psychoactive substances in therapeutic applications. Fielding's work is helping to shape the future of drug regulation worldwide, and she has been called upon by both the president of Guatemala and the Jamaican Minister of Justice to advise on their respective countries' drug policies.

A CHANGING VIEW

Though many people still instinctively oppose drug use of any kind, the overall public opinion of psychedelics is undeniably becoming more accepting. This is largely thanks to the recent decriminalization movements of cannabis, which have helped to lift the taboo against naturally-occurring psychoactive compounds. With both Uruguay and Canada fully legalizing cannabis in the last decade, more and more people are starting to find that the ingrained belief that drug use automatically leads to immorality is nothing more than an institutional myth. The cannabis legalization movement in the U.S. is progressing with strength, with the decriminalization of mushrooms riding on its heels.

Another contributor to our collectively changing perspective on psychedelics has been internet access. A quick search of "magic mushrooms" results in several high-profile articles covering recent research, as well as a handful of advocacy groups that are serving to destigmatize psilocybin and publicize its benefits.

Without diving too deep into politics, we will say simply, people have been lied to about drugs for decades. Starting with the criminalization of cannabis in the earlier part of the 20th century, governments globe-wide have increasingly tightened their grips on our ability to experiment with and alter our consciousnesses. Often, these moves have been made to suppress certain minorities, such as with the outlawing of cannabis whose users were primarily black and Latino. Otherwise, the motives for these prohibitions are largely unknown, leaving their bases to assumption. With psychedelics, the most easily-assumed motivation for their prohibition is that the authorities wish to keep the populace in a state of complacency. Psychedelics, mind-opening agents that they are, often cause users to question the status quo, thus upsetting and unbalancing the powers that be.

As time marches on, the societal injustices that continue to unfold, especially income inequality, have caused many more people to question governments' policies in several spheres of operation. Rising among the ranks of these challenges is the issue of dissonance between authoritative claims of drug dangers and the research and experiences predicating certain drugs' relative safety and benedictions.

In short, people are waking up. We are becoming aware of the fact that much of government policy is based upon spurious claims with dubious justifications. We are realizing that we have been robbed of our right to the free and unhindered exploration and control of consciousness. And, we are finding that the bureaucratically-endorsed pharmaceuticals may not be the most effective drugs for all our ailments, being effectively reminded that, for many problems, nature holds the cure.

LEADERS OF MODERN RESEARCH

Though psilocybin and psilocin remain strictly-controlled Schedule I substances under the United Nations Convention on Psychotropic Substances, the last twenty years have seen a major influx of research into their psychotherapeutic uses. Many organizations and universities have begun psilocybin trials and have published several studies showing that the original hunch of Valentina Pavlovna Guercken to be truly prescient- that is, psilocybin does indeed have highly valuable medicinal applications.

Psilocybin's range of therapeutic uses still has yet to be fully examined, but more research is happening each day with extremely heartening results. Several countries are hosts to institutions that are seeking to further our knowledge of psychedelic therapy, including the United States, the United Kingdom, Switzerland, Germany, and Canada. These organizations include universities and associations dedicated to psychedelic and consciousness research. Notable among them, those at the forefront of activity, are the Johns Hopkins University Center for Psychedelic and Consciousness Research, the Multidisciplinary Association for Psychedelic Studies (MAPS), the Heffter Research Institute, and the Beckley Foundation.

JOHNS HOPKINS CENTER FOR PSYCHEDELICS AND CONSCIOUSNESS RESEARCH

The Center for Psychedelics and Consciousness Research is the dream-come-true of founding director Roland R. Griffiths. Griffiths, a psychopharmacologist, has studied the effects of mood-altering drugs since the mid-1970s. After taking up a meditation practice in the 1990s, Griffiths became interested in the spiritual implications of psychedelic use. He met hippie-era researcher Bill Richards sometime after, and the two decided to launch an official research project into the effects of psilocybin.

The team's research began in 1999 and was the first study of its kind to gain approval from the FDA. Their findings were published in the August 2006 issue of *Psychopharmacology*, titled "Psilocybin can occasion mystical-type experiences having substantial and sustained personal meaning and spiritual significance."[88] This study, a sort of confirmation of the 1960's Marsh Chapel Experiment, aimed to precisely describe the effects of psilocybin. It found that psilocybin does indeed produce spiritual experiences with a high rate of probability, with 67% of participants rating the mushroom trip "among the top five most meaningful experiences of their lifetime". A fourteen-month follow up report found that many of the subjects (64%) had prolonged positive changes in their moods, behaviors, and attitudes and increased reports of life-satisfaction and well-being.

Griffiths continued studying psilocybin and other psychedelics over the course of the next two decades and in 2019 announced the opening of the Johns Hopkins Center for Psychedelics and Consciousness Research.[89] The center, funded by $17 million in private donations, is leading the way in psychedelic studies, focusing not only on their psychotherapeutic benefits, but also on their general effects on consciousness, perception, and neuropharmacology. Thanks to the rigorous efforts of Griffiths and his team of researchers, psilocybin is gaining ground as a safe and effective medicine for a variety of ailments and disorders.

Multidisciplinary Association for Psychedelic Studies (MAPS)

MAPS was founded by Rick Doblin in 1986 and has been a powerful force in the funding of research into psychedelics and psychedelic-assisted psychotherapy, as well as in advocating for decreased restrictions of these substances in medical research and applications.[90] Rick Doblin, a graduate of Harvard, conducted the follow-up studies to Leary's 1960s research- both the Concord Prison Experiment follow-up and the reassessment of the Marsh Chapel Experiment. He founded MAPS as a 501(c)(3) non-profit research and educational organization as a means of continuing research into the psychotherapeutic uses of MDMA, though subsequent studies would expand to look at the effects of LSD, ibogaine, and cannabis.

Though MAPS does not focus on psilocybin research, the organization has donated more than $12 million toward psychedelic studies since its 1986 inception. With a worldwide reach, they continually support research in the U.S., Canada, Israel, Jordan, and Switzerland.

Heffter Research Institute

The Heffter Research Institute was founded in 1993 by pharmacologist and medicinal chemist David E. Nichols with co-founders Dennis McKenna, Charles Grob, Mark Geyer, and George Greer.[91] While the institute's early research focused on MDMA and ketamine, the focus became psilocybin starting in the early 2000s. The Heffter Research Institute funded the first clinical research study since the 1970s in the University of Arizona Medical School's study titled "Safety, Tolerability, and Efficacy of Psilocybin in 9 Patients With Obsessive-Compulsive Disorder".

The institute has further funded studies examining the efficacy of psilocybin in reducing cancer-related stress and treating alcohol and nicotine addiction. Additionally, it has funded research into the spiritual and neuroscientific implications of psilocybin. Much of this research into the neuroscience of psilocybin takes place at the Heffter Research Center at the University of Zurich in Switzerland.

The Beckley Foundation

Drug policy reformer and research coordinator Amanda Fielding created the Foundation to Further

Consciousness in 1998, which was later renamed to the Beckley Foundation. The foundation's stated goals are to scientifically investigate the effects of psychoactive substances on the brain and consciousness, learn more about consciousness and brain function, discover and explore new avenues for the treatment of illnesses, and to achieve rational, evidence-based changes in global drug policies. Partnering with scientists and research institutions, the Beckley Foundation initiates, supports, and directs psychedelic research projects.[92]

RECENT STUDIES INTO PSILOCYBIN'S POTENTIAL

Research into the medicinal and therapeutic applications of psilocybin is being conducted in several different areas. These range from mental health treatments for depression, post-traumatic stress disorder, obsessive-compulsive disorder, and anxiety, to means of expanding creativity and empathy, to crime reduction, addiction treatments, cluster headache treatments, and easing end-of-life distress. There have also been studies done assessing the safety and risks associated with magic mushroom use, generally resulting in the conclusion that they are one of the safest drugs in popular circulation.[93]

The results of these studies are uplifting to those of us who support the decriminalization of magic mushrooms, as they are continuously showing that psilocybin is a safe, effective drug with overall positive outcomes in a vast array of uses. As research continues, we will undoubtedly learn more of its medicinal and cultural value, hopefully concluding in the lifting of restrictions imposed by the world's governments and a reclassification from the non-medical Schedule I to a more appropriate scheduling. What follows is a brief outline of a few of the ways psilocybin can be used to treat some of society's major ailments.

Depression

Depression manifests in many people worldwide, with an estimated 3% of the global population thought to be affected by major depressive disorder. Conventional depression medications, those pharmaceuticals currently approved for prescription, are effective up to a point but have been called into question regarding their tendency to increase the prevalence of suicidal ideations, as well as for falling below the margin of effectiveness compared with placebos. Some researchers have declared current SSRI treatment to be clinically insignificant.[94]

Psilocybin therapy for the treatment of depression has returned great results, finding that after even one dose, subjects report a significant improvement in depression symptoms for up to five weeks. A 2015 study by the University of Alabama concluded that psychedelic use may prevent suicide, significantly reducing suicidal ideation and planning and reports of psychological distress.[95] Clinical trials conducted by the Imperial College London found that after two psilocybin treatments, patients with treatment-resistant depression had rapid improvement of symptoms that persisted for up to six months following the administration of the drug.[96] Several other studies have been done on using psilocybin as a treatment for depression and related anxiety disorders, and many more are ongoing.

Quite recently, the FDA has granted breakthrough therapy status for psilocybin therapy to two U.S. companies utilizing the power of magic mushrooms to treat depression.[97] This designation as a breakthrough therapy is given to drugs that show substantially-improved results as opposed to the currently available treatments. It is a significant change of attitude, demonstrating that the official opinion toward psilocybin is gradually improving, and may soon lead to its legal rescheduling.

Post-Traumatic Stress Disorder

PTSD affects even more people than depression, with approximately 8% of the global population having PTSD at some point in their lives.[98] Drugs currently used to treat PTSD are hardly more effective than current depression medications, improving symptoms to only low or moderate degrees.[99] Psilocybin-assisted psychotherapy is being investigated as a possible treatment, and along with MDMA, is showing markedly effective results.

Research in this area pertaining to psilocybin is a bit more limited, as MDMA is a clear frontrunner for PTSD treatment, having been given breakthrough therapy status by the FDA in 2017. However, a study in 2013 found that psilocybin administered to mice can induce neurogenesis of the hippocampus and cause an extinction of the fear response.[100] The neurogenesis findings are remarkable in themselves, and additionally, the extinction of the fear response shows that the trauma behind PTSD may simply be eradicated following psilocybin administration.

A human trial in 2012 using fMRI scans revealed that psilocybin reduces activity in the medial frontal cortex of the brain.[101] Overactivity of this region is one of the markers of PTSD symptoms. The results of both studies show psilocybin may be very helpful in treating PTSD in conjunction with psychotherapy. In light of the research showing psilocybin's treatment potential in treating depression and PTSD, a psilocybin nasal spray has recently been developed by the U.S. company Silo Wellness in anticipation of an imminent decriminalization of the magic mushroom molecule.

End-of-Life Distress

Among terminal cancer patients are a high prevalence of depression, anxiety, and reduced quality of life related to the foreboding of approaching death. With this comes an increased rate of suicide, for which traditional therapy and medication have largely been ineffective.

In recent clinical trials carried out separately by researchers at Johns Hopkins University and New York University Langone Medical Center,[102,103] cancer patients administered psilocybin showed marked improvement in depression and anxiety symptoms, along with improvements to their quality of life. In the NYU study, there was a one-time dose of 0.3 milligrams per kilogram of body weight, whereas in the Johns Hopkins study, patients were first administered a small dose of one to three milligrams per kilogram followed by a higher dose of twenty-two milligrams per kilogram around five weeks later.

The results of both studies were similar, with positive effects lasting for up to eight months after the psilocybin sessions. Quality of life improvements included higher levels of energy, greater amounts of socialization, improved relationships with family members, and better job performance. Also reported were feelings of unexpected peace, altruism, and spirituality. These trials, supported by other studies which show a general improvement in mood and quality of life, show that psilocybin can be an effective means of comforting those suffering from end-of-life distress.

Addiction

Addiction is an ailment that has plagued humanity for millennia. Despite various treatment options, many addicts ultimately return to their addictive behaviors shortly after the conclusion of their treatment. Contrary to being addictive, psilocybin has been shown anecdotally and experimentally to aid in the cessation of addictive substances and behaviors.

In a 2017 survey analysis of 44,000 illicit opioid users,[104] the Laboratory for Integrative Psychiatry at the McLean Hospital of Harvard found that psychedelic use among illicit opioid users is associated with a 27% reduced risk of dependence and a 40% reduced risk of abuse. A small proof-of-concept clinical trial conducted by the University of New Mexico Health Sciences Center found psilocybin administration in alcoholics to be connected with a reduction in cravings, changes in drinking habits, and increased abstinence self-efficacy up to 36 weeks after the initial treatment.[105]

A highly successful trial at Johns Hopkins University studied the effects of psilocybin in smoking cessation.[106] Fifteen participants were given three doses of psilocybin, with the first dose administered on the day the subjects planned to quit smoking, and the following two doses occurring at two and eight weeks later. Six months after the trial, 80% of participants had remained smoke-free. In comparison, the drug varenicline, which is currently considered the most effective smoking-cessation drug, has an approximate 35% six-month success rate. Another study on psilocybin-assisted smoking cessation is currently underway at Johns Hopkins.

Crime Reduction

The connection between psychedelics and criminal tendencies is currently only tenuous, but two studies have shown a correlation between psychedelic use and an inhibition of criminal behavior. The research, both joint ventures by the University of Alabama Birmingham and Johns Hopkins University School of Medicine, analyzed data regarding drug use and crime. In the first study,[107] researchers gathered information on over 25,000 people enrolled in the Treatment Accountability for Safer

Communities program (TASC). TASC is a community corrections supervision program for those with a history of drug abuse. Of the 25,622 participants, 1% were reported to have a hallucinogen abuse disorder. Those reporting hallucinogen use were found to be less likely to fail the requirements of the TASC program, less likely to be incarcerated, and more likely to appear for appointed court dates. The study concluded that hallucinogen use may predict abstinence from alcohol and other illicit substances and promote prosocial behavior.

The second study analyzed thirteen years of data from the US National Survey on Drug Use and Health, encompassing information on over 480,000 individuals.[108] Those with a reported use of classic psychedelics at least once in their lifetime (DMT, mescaline, ayahuasca, LSD, peyote, psilocybin) were found to be 27% less likely to have committed theft, 12% less likely to have committed assault, and to have a 22% decreased odds for past year arrest for property crime and an 18% decreased odds for past year arrest for violent crimes.

These studies in themselves are not enough to establish a firm link between psilocybin and crime reduction, but they do point in that direction. Furthermore, a study from Maastricht University in the Netherlands found that a single dose of psilocybin can increase empathy for at least seven days.[109] As criminal behavior is strongly correlated with a reduced empathy competence,[110] the connection between psilocybin and a reduced likelihood of criminal behavior is becoming stronger as the body of evidence grows.

PSILOCYBIN AND THE BRAIN

We are still learning about how exactly psilocybin works-
that is, how it affects brain chemistry, its mechanisms of
action, and how it makes us trip. The study of psychedelics
is hand in hand with the study of consciousness, which
we know altogether very little about. Without bogging
you down with too in-depth a look at what we know so far
about how psilocybin changes our brains, we will give a
brief outline of its known pharmacology.

Pharmacodynamics: Neurotransmitters and Mushrooms

We, the authors, are by no means a neuroscientist or
pharmacologist, but we willl try to present a simplified
version of what happens in our brains when we use sacred
mushrooms.[111] After you eat magic mushrooms, the psilocybin is quickly metabolized into psilocin. This happens
in the stomach, kidneys, intestines, and blood. After
ingestion, psilocybin and psilocin can be found in blood
plasma within twenty to forty minutes. Psilocin, being
lipid-soluble, is then distributed through the bloodstream
and across the blood-brain barrier. Studies have shown a
high probability that psilocin alone is responsible for the
psychological effects of magic mushrooms.

Psilocybin and psilocin are classified as tryptamine
alkaloids, molecularly similar to the neurotransmitter
tryptophan, and structurally similar to serotonin. Once
converted, the psilocin begins binding to and activating

various serotonin receptors, mainly the 5-HT2A receptors. At this point, science lacks a clear answer as to what happens. Several changes in neurochemistry take place, but we're not sure how exactly this causes us to trip. For one, the binding of psilocin to these serotonin receptors activates the enzyme phospholipase C. Following this is the activation of the enzyme protein kinase and a cascading release of calcium ions. Essentially, this is the jumpstart to the chemical changes that cause hallucinations.

Other neurochemical changes which may occur effectively unsettle the normal serotonin output in the brain, increasing levels and rerouting it to different neural regions. Additionally, the binding to the 5-HT2A receptors may stimulate glutamine production, causing changes in cellular function and communication, and disrupt the process by which the brain filters sensory information.

Psychedelic pharmacology is an extremely complex field, and overall we understand very little about why we trip so hard when we take certain drugs. As research continues, we're sure to learn more about exactly what happens in our brains to give us visions of God and feelings of perfect peace.

Temporary Rewiring

The Beckley Foundation and Imperial College's fMRI research has revealed that psilocybin is essentially a sort of eclectic neural electrician and plumber. In their 2011 study,[112] fMRI scans revealed that psilocybin causes reduced blood flow in the medial prefrontal cortex and the posterior cingulate cortex. These brain regions are responsible for many aspects of consciousness, but more importantly, are central "connector hubs" for the brain as a whole. The summary of this study concluded that the reduction in activity in these two regions allows a freer cognition due to the lack of constraint imposed by these areas.

In an ongoing collaboration, the two organizations released a further fMRI study in 2014. This case was set to investigate the interconnectedness of the brain during a psilocybin experience. The participants were divided into a control and an experimental group, the former receiving a saline solution and the latter being given psilocybin. The fMRI scans showed that psilocybin creates a huge network of connections in the brain between many different regions. Other connections are temporarily disconnected. A whole reorganization of pathways occurred, leading the researchers to surmise that this is in part what causes the sensation of an expanded consciousness during a mushroom trip.

The Beckley Foundation/Imperial College team and others are continuing to conduct research into the mechanisms of action and pharmacology of psilocybin and other psychedelic drugs, so while we may have a cloudy understanding at this time, the future will undoubtedly reveal how and why we trip on shrooms.

The top organizations in psychedelic studies are continuing to bring innovative and enlightening research projects to fruition. In stark contrast to the Harvard Psilocybin Project's rather poorly designed and executed studies in the 1960s, these current investigations into the effects and mechanisms of psilocybin are being carried out with rigorously-enforced scientific measures. Not only does this methodical protocol vastly improve our understanding of the drug, but it also serves to add credibility to psychedelic studies and to psilocybin as an accepted medication. Gone are the days when magic mushrooms were simply the spiritualist's ticket to enlightenment and the hippie's fast-track to bliss. Now is the time for professionalism, meticulous analysis, and thorough inquiries unhindered and untainted by hints of mysticism and sorcery. The spiritual benefits of mushrooms are inherent in their nature and always will be, but we can not allow a glimmer of paradise to chain psilocybin to earthly taboos any longer.

To this end, the leading psychedelic scientists in the world have many studies ongoing and planned. Many of these are still working to assess the fungi's therapeutic applications. The Johns Hopkins Center for Psychedelic Research is currently drafting volunteers for further research in psilocybin-assisted smoking cessation,[113] for studies analyzing its benefits in general depression and Alzheimer's-specific depression, and for a study testing the possible use of psilocybin in treating those with anorexia nervosa. Ongoing research at the Center for Psychedelic Research is further investigating the potential of psilocybin to help terminal cancer patients cope with end-of-life distress. Additionally, in homage to psilocybin's ritual roots, Johns Hopkins is recruiting religious leaders for help in studying the mystical aspects and entheogenic applications of magic mushrooms.

With the recent FDA bestowments upon psilocybin as a breakthrough therapy for depression, Compass Pathways and the Ursona Institute are moving to advanced phases of clinical trials using psilocybin for treatment-resistant depression. Part of this research is being funded by the Heffter Institute, which is furthermore working to develop other research centers at universities around the world.

The Beckley Foundation, in addition to its own examination of psilocybin as a stop-smoking aid, has partnered with the Imperial College Psychedelic Research Program to study the neuropharmacological effects of magic mushrooms. Using BOLD fMRI imaging, the team is measuring the psilocybin-induced increases and decreases of activity in different regions of the brain, as well as how the drug impacts connectivity between different brain regions. Additional neurological studies are being conducted by the Heffter Research Center at the University of Zurich, combining brain-imaging with measures of psilocybin's effects on attention, perception, and cognition. Likewise, research at the Medical School Hannover is looking into the effects of psilocybin on various areas of visual perception, which may serve to support or undermine Terence McKenna's Stoned Ape Hypothesis.

Studies of this sort were impossible in the early years of psychedelic research as our medical technology was not yet advanced enough to allow this type of imaging. Now, with the ability to analyze brain activity in real-time, we are learning more about not only the psychological effects of psilocybin but its neurological implications as well. This is greatly important because up to now, the methods of action of the drug were all but unknown. To advance it as a safe and accepted treatment for any ailment and to move toward a reclassification as a less-restricted substance, we must know as much about its mechanisms as we can.

HOW YOU CAN HELP THE MOVEMENT

Rather than sit by and watch as the events around psychedelics unfold one way or the other, there are things you can do to have an impact on the future of their legalization and medical applications. Our first piece of advice is to practice harm reduction. Please, don't get high on mushrooms (or any other psychedelic) and do something stupid. Don't trip in public. Don't get high and drive. Avoid doing anything that can bring a bad look to the already-burdened perspective on psychedelics. Trip responsibly, be upright and moral before, during, and after your drug experiences, and help to dispel the myth that drug use is directly linked to immorality, crime, and corruption.

To take a more active approach, supporting local advocacy groups is a good place to start. Search for psilocybin/
psychedelic activism organizations in your area, contact
them, and ask what you can do to help. There may be peti-
tions to sign, buttons to wear, meetings to attend- the basic
trappings of any other advocacy program.

You can also donate to research centers or volunteer to
become a study participant. You may or may not get a free
mushroom trip through by being a test subject, but either
way, you'll be contributing to science through your partic-
ipation. If you decide to donate, any amount is helpful.
Since these programs lack government funding, all current
psilocybin research depends on private investors. You can
easily check the websites of any of these organizations to
find out how to donate or volunteer as a research subject.

Maybe most importantly, you should educate yourself.
To be armed with knowledge is to have at your disposal the
greatest tool in fighting the fear and ignorance surrounding
psychedelic use. Advocacy happens at the grassroots level.
If you are able to present facts regarding the safety, efficacy,
and benefits of magic mushrooms when confronted with
voices attesting to the contrary, you will be doing a large
part in moving these amazing fungi from the dark realm of
disapproval to the stadium of mainstream acceptance.

We are fortunate enough to find ourselves in the middle
of the second great psychedelic renaissance. The authors
feel hopeful that within the next several years, we will see a
great change of opinion regarding psychotropic drug use.
People are coming to see that these substances are not just

for the drop-out beatniks and peace-preaching hippies, but are rather powerful tools to aid us in our illnesses, increase our creativity, and develop a profoundly positive collective well-being. By supporting the psychedelic cause, you may, in fact, be helping to usher in the next era of human achievement and conscious evolution.

THE BASICS OF FUNGICULTURE

Rather than buy a bunch of guide books, hike out to a cow pasture, and try your luck at finding and correctly identifying a magic mushroom, you might decide you'd like to grow them yourself. It may seem like a daunting task at first, and there is without a doubt a learning curve and process of trial and error you will have to go through in the beginning, but growing mushrooms at home is actually a fairly easy and extremely rewarding project. There are countless books you can read on the subject and a variety of techniques you can try, but here, we will lead you through a few of the most popular. But first, there are some basics we need to cover regarding general practices, mushroom biology, and cultivation.

First, we must tell you again that the authors and the publisher of this book assume no responsibility or liability for any adverse effects resulting from the use or possession of any psychotropic plant, mushroom, or other hallucinogenic substance.

Familiarize yourself with the laws in your area and grow magic mushrooms at your own risk. It is safe to assume that, unless you live in Brazil, the British Virgin Islands, the Netherlands, Austria, Samoa, or Jamaica, the cultivation of psilocybin mushrooms is illegal. Therefore, proceed with caution, with intelligence, and with your wits in any pursuit you may take to follow the methods outlined in this chapter.

With that out of the way, let's talk about how mushrooms grow.

THE LIFE-CYCLE OF A MUSHROOM

If you've got a basic understanding of mushroom biology,[114] you probably know that fungi reproduce via *spores*. Spores are microscopic cells containing all the information a fungus needs to grow. In nature, they are carried on the wind, in water, or on animals until they drop off in a suitable location. This suitable location, the growth medium, will henceforth be called the *substrate*. Once a spore has reached a substrate, if conditions are right, it will begin to grow fine threads known as *hyphae* in a process called *germination*. These hyphae expand, lengthen, branch, and combine to form *mycelium*, which is the vegetative part of a fungus. Mycelium is the truest, longest-lasting form of fungi, whereas what we call mushrooms are actually only the reproductive organs of the mycelial mass.

After mycelium has started to form, it grows exponentially through the substrate, decomposing the surrounding organic matter and absorbing nutrients. Upon reaching an appropriate size, sections of mycelium begin to clump together to form *hyphal knots*. This is the first sign of a *fruitbody*. Mycelium will continue to gather at these hyphal knots, forming what are called *primordia*, or more casually, *pinheads*. Thousands of these pinheads will form from a single mycelial mass, of which relatively few are selected for further development. At this point, many of the pinheads will stop growing, while the mycelium diverts all its energy and nutrients into the growth of the select fruitbodies. The fruitbodies, also called *fruiting bodies*, are what we think of as mushrooms- stem, cap, gills, the whole shebang. The final step in the fungal life-cycle is the growth of the fruitbodies to maturity, which then release spores and fade back into obscurity. The mycelium will continue to grow and produce fruiting bodies so long as the substrate remains suitable, and its spores will travel to start new colonies in other locations.

WHAT MAKES A HAPPY MUSHROOM?

Growing mushrooms is vastly different from growing plants, but they still need the same basic elements of life: light, water, and nutrients, just in differing amounts. Additionally, mushroom spores and mycelium are much more prone to contamination than plants, so sanitization and sterilization are very important in each of the methods we will discuss.

Unlike plants, mushrooms need relatively less light to grow, and never intense light from a direct source. They also need it at different phases of their growth. Light is what signals to the mycelium that it is time to start forming fruiting bodies, so it is important to keep your grows in the dark during the mycelial growth phase.

Watering mushrooms is also different from watering plants. Rather than fill a pot using a watering can, mushrooms receive indirect water through misting their containers or the use of humidifiers. They rely on humidity for their moisture. For many mushroom species, the ideal relative humidity levels should be over 90%, with substrate moisture levels between 50 to 75%. Temperature control is also an important factor in fungiculture and will need to be adjusted at different points during the mushroom's life-cycle.

Plants can be fertilized throughout their growth process, whereas mushrooms need all their nutrients provided in the initial substrate. There are different types of substrate you can use, including straw, coco coir, brown rice flour, and rye, each with its own particular benefits. In most cases, the nutrient component of the substrate is mixed with *vermiculite,* a mineral compound, to improve water retention.

All these factors must come together in a fungiculture project to get good yields of fruitbodies. With too much moisture, your mushrooms can rot. With too much light and at the wrong times, the mycelium can dry out or begin producing fruiting bodies before the substrate is fully inoculated. The most common problem you'll experience is contamination, which can either completely destroy your grow or cause your mushrooms to be unsafe to eat due to foreign molds and bacteria. Thankfully, many magic mushroom enthusiasts have worked to develop some tried and tested cultivation methods that dial all these factors into perfection, so there's not a lot of guesswork you'll have to do if you adhere to one of the following methods.

KEEPING IT CLEAN

Throughout this chapter, you will read the words "sterilization" and "sanitization" time and time again. This is because foreign contaminants are the number one reason mushroom grows fail. It is almost certain to happen to you at some point, and it is disheartening having to discard a jar or bag of magic mycelium, but you can take steps to reduce the rates of contamination and hopefully have better success with each grow.

For general purposes, every piece of equipment and every surface you will use in your fungiculture project should be sterilized before use. You can sterilize metal tools by, carefully, heating them with an alcohol flame or butane torch. Medical-grade disinfectant alcohol or Lysol can be used to sanitize most working surfaces, but should not be used on equipment that will come into direct contact with the mushroom materials. DO NOT use alcohol or Lysol in conjunction with an open flame, as these substances are highly combustible, and using them in such a way can result in serious injuries!

A pressure cooker is an invaluable implement for fungiculture and will help reduce your contamination levels many-fold. If you can't afford a new one, they may be easily found at secondhand shops, thrift stores, and yard sales at a steep price reduction. They are necessary in most mushroom cultivation methods for the sterilization of the substrate and jars in which the mycelium will be inoculated.

Lastly, good hand-washing procedures and general hygiene are vital throughout the process. Wearing gloves which you can change or sterilize is also a good idea. Make sure to wash your hands with antibacterial soap any time you change tasks or touch an unsterilized surface. It's also helpful to wear a hairnet or cap to avoid bacteria-tainted particles contacting your grow through unseen hair debris. These sterilization measures may seem tedious, but they will greatly increase your odds of achieving successful fruiting bodies.

DIFFERENCES OF OPINION

There are many different methods for cultivating mushrooms, and an equal number of opinions on how to best perform each technique. Farming of any sort is a lot less cut and dry than it may appear to be. There are some generally agreed-upon facts, but outside of these debates abound. One of the most controversial aspects of fungiculture is the need for incubating the inoculated substrate. To do this, you can build a simple incubation chamber using an aquarium heater and a couple plastic tubs. A quick search online will bring up many different configurations of this implement, but we won't go into more detail about it here as the authors are in the camp which posits that the substrate doesn't really need this heat-assisted incubation, at least if you're willing to wait.

When you're starting out with fungiculture, it's easy to become overwhelmed by the amount of information available, especially when you see that some techniques directly contradict others. We would advise you to read about a few different techniques and decide which is most feasible for you starting out. Do a couple test runs of this chosen technique, adjusting for error each subsequent time, and be really scientific in your initial process. Take note of temperature and humidity if you have the tools for measuring this. Keep track of the dates of inoculation, first mycelial growth, first pins, and everything else that seems pertinent. The more you note what you did and when, the better you'll become at identifying your mistakes along the way.

Once you get a few successful grows using one technique, it could be good to try another. Compare your results. Don't be afraid to experiment. Have fun, don't get frustrated, and don't give up.

SOURCING SPORES

You'll need a few different things to get started growing magic mushrooms at home, and the most important of these are obviously spores. We can't give you too much in the way of specifics on this topic, but with just a little bit of determination, we are sure you'll be able to find a way to obtain these mushroom starters.

It may be the case that spores are not included in your area's legal code, in which case they can simply be ordered online. This, of course, carries its own set of risks but is among the easiest means to get started in fungiculture. For simplicity's sake, the grow-methods we will outline will be centered on one species, *Psilocybe cubensis*. This is considered one of, if not the, easiest strain of magic mushrooms to cultivate at home. It is also a particularly potent fungus, sure to contain high levels of psilocybin and set you on your merry way to TripTown. However, the methods detailed below should work for a variety of different species.

If ordering online is not a possibility for you, you can take a spore print from a fresh mushroom cap. This can be done using plain white paper, or for a more sanitary start, by using sterilized microscope slides. Our aim in this section is to keep things as simple as possible, so we will assume you only have paper to work with. Every other piece of equipment you use throughout this process should be sterilized prior to use.

So, to take a spore print, you'll first need to cut the stipe (the stem) of the mushroom off as close to the gills as possible. Next, place the cap gills-side-down on the sheet of paper. At this point, you can apply a couple drops of sterilized water to the top of the cap to speed up the spore release, but it is not necessary. All you need to do now is to cover the cap with a cup or bowl and wait. In about twenty-four hours, you can remove the cap and will see that it has left behind a dark, round stain on the paper. This is your spore print, containing millions of spores now ready to be germinated. Fold the paper and seal it in an envelope to prevent contamination, and that's that.

However, you may not have access to a fresh specimen from which to take a print. If this is the case, and you also can't order online, your last option is to attempt a culture using *agar*. Agar, derived from red algae, is a gelatinous substance often used for fungal cultivation. The two main types used in fungiculture are Potato Dextrose Agar (PDA) and Malt Extract Agar (MEA). For this means of obtaining your magic mushroom starter material, you will need a dried mushroom cap, PDA or MEA, and a sterilized glass dish, such as a small jar or a Petri dish. Simply fill the dishes with the sterilized agar, and holding the cap of the mushroom just above the dish, tap the top of the cap several times in hopes of knocking loose some spores. Next, break or cut off a couple gills and place them in each quadrant of the dish. Cover the dish and wait. If successful, you will begin to see mycelial growth within seven days, with full inoculation achieved within twenty days. This is a hit-or-miss scenario that doesn't always result in a successful mycelial cultivation, but is better than nothing.

By far, the simplest and easiest way to start your mushroom grow is by ordering spores. They come pre-sterilized, shipped in easy-to-use syringes, and are strain-specific to ensure you get a good yield. Plus, they can be stored in a ziplock bag in the fridge for up to five months. Take a look at the laws in your area concerning the purchase of spores, and if it's not strictly forbidden, this is the route I'd recommend. Now, let's get into the techniques.

THE BASIC MCKENNA METHOD

This technique is adapted from the 1970s guide written by the McKenna brothers.[115] Our presentation of this method is adapted for those wishing to put it the bare minimum effort. It goes without saying, you'll get in what you put out, so this essentials-only adaptation will yield mixed results, though it should be adequate to give you a sense of what it means to grow your own magic mushrooms.

MATERIALS YOU'LL NEED

Equipment

- pressure cooker and stove (for sterilization)
- wide-mouth quart-size Mason jars with dome and ring lids
- spray bottle
- alcohol flame lamp (preferable) or butane torch (for sterilization)

- aquarium, terrarium, or other container for storing the jars

"Ingredients"

- magic mushroom spores
- human food-grade rye berries
- calcium carbonate, such as powdered oyster shell, powdered chalk, or powdered limestone (optional)
- soil, such as a mixture of peat moss, vermiculite, and sand

THE METHOD

Compared to other methods, this is a very simple way to grow psilocybin mushrooms at home. The hardest part of this process will be preventing contamination by rigorous sterilization and sanitary measures. Some jars are almost certain to be contaminated, and these should have their contents discarded (away from your clean jars!), then be thoroughly cleaned and re-sanitized.

Jar and Substrate Sterilization

The first step of this method involves the inoculation of the rye substrate. First, you must sanitize your jars. You can do this before adding the rye as a further measure against contaminants, but let's suppose you want to kill two birds with one stone. Your Mason jars should at least be clean. To these clean jars, add 150 grams (160 milliliters) of food-grade rye berries. It is important to use rye meant for human consumption, as rye sold for animal feed is often treated with antifungal chemicals. Next, add 130 milliliters (4.4 fluid ounces) of tap or distilled water to each jar. At this point, you can add the calcium carbonate as a preventative step against foreign molds, but it is an optional additive and can be skipped.

Once your jars contain the rye and water, screw the lids on loosely with the rubber seals facing up. Now, it is time to sterilize them. You should be familiar with the mechanisms of the pressure cooker, including the pressure gauge and the stopcock valve. Pour 1.5 liters of water into the pressure cooker and add the jars of rye. If the pressure cooker is large enough, you can stack jars two rows high to sanitize more at one time. Leaving the stopcock valve open, seal the pressure cooker's lid and place it on the stove at high heat. Keep the valve open until steam begins to pour out, and then close the stopcock. Watch the pressure gauge, and when it reads 15 to 20 pounds of pressure, reduce the heat. You want to maintain 15 to 20 pounds of pressure for one hour.

After an hour at this pressure, turn off the stove, and allow the pressure to decrease to zero pounds. At this point, open the stopcock, wait for the steam to dissipate, then remove the lid. Being mindful of the heat, remove one jar at a time, tighten the lids (but don't overtighten), and examine each for any cracks or damage. Any jars showing signs of damage should be discarded, as cracks will allow foreign molds to contaminate the substrate. You may wish to write the date on the jars in order to track the progress of your grow. As you examine each jar, shake it vigorously, and then place it on a pre-sterilized working area, allowing them to cool to room temperature.

Inoculation

Now, with the jars at room temperature, you are ready to inoculate your rye substrate with spores. If you used an agar method to germinate your spores, use a sterilized blade or scalpel to slice the agar medium into small squares of approximately 1 centimeter in size. Resterilize the blade, spear an agar square, and then quickly place it into a jar by cracking the lid just wide enough to fit the blade. Close the lid, shake it vigorously, and repeat for each jar. If you are using a spore syringe, the process is much the same. Simply sterilize the needle by flaming it with the alcohol lamp or butane torch. Extinguish the flame, then wipe the needle with an isopropyl alcohol swab. Insert between the lip of the jar and the barely lifted lid, and inject roughly between 0.5 cc's to 1 cc of solution into the substrate. Close, shake, repeat.

Colonization

Following the inoculation of all jars, set them with loosely-sealed lids into your aquarium or other container. The temperature in the grow area should be maintained between 70F to 80F. You can check on them every few days, but not much will seem to happen at first. Starting on the fourth day after inoculation, begin a process of examining and shaking each jar. With clean hands, remove a jar from the group and firmly tighten the lid. Check for any signs of mold and mildew, which will often appear as blue-green crusts, yellow slimes, or various black, green, or grey molds. If the jar appears contaminant-free, shake it well to scatter the mycelium throughout the substrate. After the final shake, loosen the lid just a crack and smell. If it smells of anything other than cooked barley, it is probably contaminated and should be discarded. Take care to remove all contaminated jars and to clean them well away from your grow area to avoid spreading the contaminants to your clean samples. Jars with foreign molds can not be salvaged, so though it may hurt a bit to toss out your psilocybin mycelium, it must be done for the benefit of the whole of the project.

Within one to three weeks, the mycelium should be completely spread throughout the rye substrate. The sides of the jar will clearly show a snow-white webwork of mycelial strands, sometimes with a light blue tinging. If you have made it this far, you are just one step away from attaining fruitbodies!

Casing

Finally is the step called *casing*. Casing involves the application of sterile soil to the top of the rye substrate. If you are using commercial soil, it may not be necessary to sterilize it, but if you wish to take the added precaution against contaminants, you can sterilize it in jars using the pressure cooker in the same manner that the rye was sterilized. Before sterilizing, and before application to the substrate jars, the soil should be wetted. This can be done by spreading it in a thin layer on a clean sheet of plastic and misting it with a garden hose. Alternatively, small amounts of soil can be wetted in a mixing bowl using a spray bottle while stirring with a spoon. The goal is to moisten the soil without saturating it. Once it is wet, you can either sterilize it or begin casing your jars.

To case your inoculated substrate, remove the lids of your jars and add approximately ½ cup of pre-wetted soil to each. Lightly shake the jars to evenly distribute the soil, and then further wet it with a few mists from a spray bottle. The jars should then be stored in the container or aquarium with the lid of the container partially removed. You want to maintain the temperature above 70F and ensure a minimal amount of air movement.

Watering now becomes a daily practice. Each jar should receive two or three pumps from the spray bottle per day. If your area's climate is particularly hot and dry, more may be needed. Likewise, if conditions are very humid, less water should be applied. The key is to keep the casing soil moist and spongy, avoiding dryness or dripping saturation.

Fruiting, Harvesting, and Re-casing

Between two and three weeks, you will begin to see mycelium growing through the casing soil. Next will come the pinning stage, in which you will notice small clumps of mycelium around the edges of the soil. Some of these will soon start to form into distinct mini mushrooms. After these pinheads break the surface of the casing soil, they will grow to maturity within five to ten days. There will be aborted primordia along the way, which can also be picked and eaten.

Once the first flush of mushrooms reaches maturity, you can pick them and either eat them fresh or dry them. Fresh mushrooms can be stored in a plastic bag in the refrigerator for seven to ten days. To dry the mushrooms, you can construct a box specifically for this purpose, use a dehydrator, or simply place them on racks with a stream of air from a fan directed onto them. They should be completely dried before storage, being brittle to the touch with no sponginess whatsoever. When they are dry, they can be sealed in plastic bags and frozen to preserve freshness and potency.

And then you've done it! The jars can then be recased, by using a sterilized fork for each jar to remove the previous casing, clean out the aborted mushrooms and pins attached to the side of the jar, then replacing the old casing and covering it with another ½ inch of pre-moistened soil. The fork should then be used to gently push the fresh soil into the space between the old casing and the jar sides. With proper watering, the next flush should appear within a few weeks. On average, you can expect each jar to fruit for a total of between 60 to 80 days, yielding roughly 10 grams of mushrooms per every 100 grams of rye. Just remember to keep them hydrated, aerated, and contaminant-free, and you'll soon have mushrooms for months.

PF-TEK

The PF-Tek method is a very popular way to grow mushrooms at home in a low-maintenance, low-cost setup.[116] This technique (tek) was devised by Robert McPherson under the pseudonym Psilocybe Fanaticus, hence the name PF-Tek. It is considered by many to be the easiest technique for beginners and can yield up to 200 grams per month with a single fruiting chamber. Some have criticized PF-Tek as having a low average of success, but it is still a favored method employed by many novice and experienced fungiculturalists. There are several variations on the technique, and it has been updated as more growers have improved upon Robert McPherson's original scheme. Below, we present a PF-Tek variant that has shown high success rates while keeping necessary materials and experience to a minimum.

MATERIALS YOU'LL NEED

Equipment

- pressure cooker and stove (for sterilization)

- wide-mouth half-pint Mason jars with dome and ring lids
- professional-grade masking tape
- ice pick or nail
- aluminum foil
- spray bottle
- alcohol flame lamp (preferable) or butane torch (for sterilizing the syringe)
- clear plastic bin with lid (large enough to fit all your jars with a few centimeters spacing)
- electric drill with a 6 millimeter and a 3 millimeter bit
- 5,000-6,000 Kelvin "natural daylight" fluorescent light and ballast (optional)
- bricks, or something similar (to use as a stand for the fruiting chamber)
- plastic sheeting or tarp (optional, to catch excess water dripping from the fruiting chamber)

"Ingredients"

- mushroom spore syringe
- brown rice flour (can be bought as flour or you can grind your own brown rice)
- medium-grade agricultural vermiculite
- perlite

THE METHOD

Jar Preparation

The first thing you will need to do for the PF-Tek method is to prepare your jar lids. You will need to create two holes in each jar lid using the ice pick or nail. Orient the dome part of the lids with the rubber seals facing up, and create two punctures at opposite ends of the lid, right past the inside edge of the sealing area. If you use a nail for this, you can hold it in a pair of pliers to maximize your leverage. Use the masking tape to attach the dome of each lid setup to the ring part, so both pieces function as one whole, keeping the dome oriented so the rubber seal faces up when placed on the jar. Cover the holes with masking tape to prevent contamination during the inoculation phase.

Substrate Preparation

Next, you will prepare your substrate. For six half-pint jars, you'll use 1 liter of vermiculite, 0.5 liters of water, and 0.5 liters of brown rice flour. This comes out to approximately 160 milliliters of vermiculite (0.7 U.S. cup) and 80 milliliters (around 1/3 cup) of brown rice flour per jar. Start by slowly adding water to the vermiculite in a large mixing bowl. Stir with a large spoon as you add water, and aim for saturation. Periodically press the spoon into the vermiculite. Once a small amount of water starts to issue forth, you have added enough water. Make sure it is well-mixed, then mix in your brown rice flour. It is important to do so in this order or the substrate will achieve the wrong consistency.

Once the substrate is prepared, you will add it to the jars. Spoon in the substrate mixture until each jar is filled to 1 centimeter from the top. Do not tamp the mixture down as you add it, though you can lightly shake the jars to close any large air pockets. Fill the remaining 1 centimeter of space with dry vermiculite, then place close the jars with the prepared lids. Cover each jar lid with a piece of aluminum foil to prevent water from dripping onto the dry vermiculite.

Sterilization

Next, you will sterilize the jars in the same way as the McKenna method. Put 1.5 liters of water in the pressure cooker, set the jars in the bottom (using a steaming rack is recommended to prevent breakage). Seal the lid, keeping the stopcock open, and follow the procedure outlined in the first grow method. Since you are using smaller jars, you can reduce the cooking time to one hour. As in the McKenna method, allow the cooker's pressure to recede to zero before removing the jars, and leave them to cool to room temperature.

Inoculation

Now, it is time to inoculate the jars. First, you should sterilize the spore syringe needle. It is best to do this with an alcohol flame lamp, but a butane torch can be used if necessary. After flaming the needle, wipe it with isopropyl alcohol. Shake the syringe to redistribute the spores through the solution. You will then remove one of the pieces of tape covering the holes in the lid. Insert the needle through the hole, push it in past the layer of dry vermiculite, and carefully angle the syringe so that the needle is against the side of the jar. Slowly inject 0.5 cc's of spore solution, remove the needle, sterilize, and repeat for the next hole. Do this for all jars, using 1 cc per jar of substrate.

Colonization

After all jars are inoculated, place them on a shelf or other clean surface and wait. As in the McKenna method, you should regularly inspect the jars for signs of contamination. You do not need to shake the jars or do anything further to them at this point. Simply watch for signs of mycelial growth and foreign invaders, aiming to keep the temperature in the room between 70F and 80F. It will probably take two to three weeks for the substrate to become fully taken by the mycelium. During this time, you can start to prepare the fruiting chamber.

Preparing the Fruiting Chamber

In this step, you will use the plastic bin and the drill. Using the 3mm bit, drill a series of holes approximately 2 cm apart on the bottom side of the bin and on the lower 12 cm of all other sides. Use the 6mm bit to drill holes 5 centimeters apart on all the remaining space, including the lid. If you will be using a light in the fruiting stage, now is a great time to get it set up in your growing area. Remember, indirect light is optimal and direct light can harm your grow, so try to set the light up off to the side of where you will keep the fruiting chamber.

The fruiting chamber will also need a layer of perlite in the bottom, but this step should not be done until your substrate jars are fully colonized and ready to be placed in the fruiting chamber. When it comes time to add the perlite, start by washing it in a colander. Depending on the size of your plastic bin, you may need a rather large bag of perlite to fill the bottom. After it is washed, allow the excess water to drain out, then add it to the fruiting chamber. Keep in mind, this is not to be done until the day your brown rice substrate cakes are ready to be transferred.

Consolidation and Preparation for Fruiting

After a couple to a few weeks, the substrate should be totally permeated by mycelium. Every visible part of the jar should show the thick, snow-white of hyphae. When the jars have reached this point, label them with the date and allow them to sit for one more week. This is called the consolidation period and allows the mycelium to further form a solid cake. If during this time you see any pins beginning to form, you can remove those cakes from the jars and continue to the next step.

When the consolidation week is finished, remove the cakes from the jars, remembering to wash your hands first. Gently scrape off the layer of dry vermiculite. Next, you will submerge them under clean water for 24 hours. This serves to rehydrate them and prepare them for the next step. They will float in the water, so it is advised to rest a sterilized plate or something similar on top of them to keep them submerged. After soaking for 24 hours, rinse each cake under cool water. Then, proceed to gently roll them around in dry vermiculite. Your aim here is to have every surface of the cake coated in an even layer of vermiculite.

Fruiting

Finally, you are ready to move them to the fruiting stage. First, you should add 2 cm of dry perlite to the bottom of the fruiting chamber. Next, wash the perlite as described in the "Preparing the Fruiting Chamber" section, and add this wet perlite on top of the dry to a depth of 7-12 cm. Now, you can place the substrate cakes into the plastic bin. Space them a couple centimeters apart and place the lid on the bin. If you have more cakes than will fit in a single layer, stacking them is fine and may result in larger fruitbodies.

Artificial lighting is optional at this point. If you use a light, do not aim it directly at the fruiting chamber or the humidity levels will become too low. You can rely on natural sunlight if you wish, but some form of light is necessary at this stage. The ideal lighting routine is a 12-hour cycle- 12 hours on, 12 hours off.

The temperature should be kept between 68F and 80F, with the best growth occurring closer to 80F and extremely slow growth at the lower temperature range. During the fruiting phase, daily watering is essential, and you should try to maintain relative humidity levels at 95%. Using a room humidifier helps a lot if you live in a dry environment. Use a spray bottle to mist the cakes three or four times a day, indirectly for the first few days to avoid displacing the vermiculite layer.

Within a week or two, you will start to see baby mushrooms. Continue misting to maintain humidity, and they will soon approach maturity. A mushroom is mature when the partial veil breaks, exposing the gills. When the largest mushrooms have reached maturity, you can harvest every mushroom from the cake, as the smallest ones will not grow much bigger. To pick them, grasp them at the base and gently twist to break them off. As in the above McKenna method, you can choose to eat them fresh or dry them using your desired method.

Preparing the Cakes for Future Flushes

After harvest, each cake can be reused up to about four times. To prepare them for the next round of fruiting, remove them from the fruiting chamber and set them on a sterilized surface. Allow them to dry for a full 24 hours, and then soak them as described above. The subsequent soaking periods can be reduced to 12 hours. Repeat the washing and rolling in vermiculite steps, then place them back into the fruiting chamber. Water, wait, and repeat.

The PF-Tek method generally has the first flush as the fullest. Each flush after the first will yield fewer mushrooms, and after more than a few fruitings, the quality of the cakes will degrade and become unviable. With PF-Tek, you can expect up to 100 grams of fresh mushrooms per fruiting chamber. It may be a fair amount of work, but the rewards are well worth the effort!

SPAWN BAGS

This method is less popular with magic mushroom cultivators but is a fairly common practice with those growing gourmet edible mushrooms. It has its own pros and cons, the biggest pro being that it is probably the least labor-intensive and all-around simplest way to grow mushrooms, assuming you decide to fruit in the bag. We would say the biggest con is that you have less control overall and typically will grow only a couple bags at a time, so you leave yourself open to losing the whole project to contamination. In this section, we will focus on the fruit-in-bag technique, as well as discuss how you can transfer your mycelium-filled substrate to a fruiting chamber.

A spawn bag is the simple name for a gusseted autoclavable polypropylene filter patch bag. "Gusseted" refers to the way this bag is folded, while autoclavable refers to its ability to be sterilized in a medical autoclave. This is an important feature, allowing you to put a spawn bag in a pressure cooker like you would the Mason jars in the previous methods. Polypropylene is the type of plastic most spawn bags are made of, and the filter patch is what allows these bags to "breathe", meaning your mushrooms can be aerated as they grow.

Some companies sell bags that come pre-filled with substrate, so all you have to do is inject spores and keep them moisturized. These are a great option for those who want to put in as little effort as possible, but you may achieve better results and will definitely learn more about fungiculture if you start with empty bags and fill them with your own substrate. For informational purposes, we'll explain this method assuming you want to start from scratch.

MATERIALS YOU'LL NEED

Equipment

- 2x medium-size spawn bags
- pressure cooker and stove
- alcohol flame lamp (preferable) or butane torch
- professional-grade masking tape
- kitchen plate that can fit in your pressure cooker

"Ingredients"

- mushroom spore syringe
- brown rice flour
- coco coir
- used coffee grounds
- medium-grade agricultural vermiculite

THE METHOD

Substrate Preparation

There are many different spawn bag techniques, and this is a simplified adaptation of one of our favorites. For this method, the first step is to prepare your substrate. So, first, mix 1 liter (4 cups) of vermiculite with water, following the same procedure outlined in the PF-Tek method and aiming for the same moisture level. Once the vermiculite is properly wetted, add 0.5 liters (about 2 cups) of used coffee grounds, and mix it in well. Next, stir in 1.25 liters (5 cups) of coco coir. The last addition to the substrate is 1.25 liters (5 cups) of brown rice flour. Make sure to add this last to achieve the right consistency. Thoroughly mix the substrate.

Substrate and Spawn Bag Sterilization

This is enough substrate to fill two medium-size spawn bags. You can adjust your measurements accordingly if you are going to use more or fewer bags, but two is a good place to start for beginners. Medium-size spawn bags are those with measurements of approximately 33cm x 23cm. Spawn bags have different filter ratings, and you will want to use bags with a filter size of 0.5 microns or less. Any larger will leave your substrate vulnerable to contaminants.

After all your substrate ingredients are mixed, you can split it between the two bags. The substrate should fill them to <u>below</u> the filter patch. If you for some reason have enough substrate to fill them above the filter patch, remove some of the matter. This is crucial to allow your bags to be aerated. When the correct level of substrate is in the bags, press the air out, fold the opening, and tape it shut with a piece of masking tape. Fold it again and apply another piece of tape. Make sure that the bags are sealed well. Next, put two pieces of tape in different spots on the side of the bags opposite the filter patch. These will be your injection points.

Now that the bags are closed, you can put them in the pressure cooker. As in the previous two methods, you want 1.5 liters of water in the pressure cooker, but you must be sure to keep the filter patch above the water level. Using jar lids or a combination of these with a steamer basket to elevate the bags from the bottom of the cooker is a good idea. Orient the bags with the opening toward the top, then follow the pressure cooker sterilization instructions outlined in the McKenna method. Two to two and a half hours is an appropriate amount of cooking time for two bags. After this period, let the pressure cooker return to equalized pressure.

Inoculation

Remove the bags from the pressure cooker, and place them on a sterilized surface to cool to room temperature. This may take up to 12 hours. Once they are cool, you can move to the inoculation step. First, wipe the tape patches on the back of one of the bags with isopropyl alcohol. Shake the syringe to redistribute the spores through the solution. Then, flame your syringe needle and wipe it with alcohol as well. Insert the needle through the tape about 1 cm into the substrate, and inject 0.5 cc's of spore solution. Angle it to another point in the substrate and inject another 0.5 cc's, being careful not to poke another hole in the bag. Withdraw the needle from the bag, and apply a clean piece of masking tape over the injection site. Repeat the steps for sterilizing both the tape patch and the needle for the remaining three injection sites, and inoculate each bag with a total of 2 cc's of spore solution. Remember to cover each punctured injection site with a fresh piece of tape as soon as you remove the needle.

Colonization

Like with the PF-Tek and McKenna methods, now comes the waiting phase. Store the bags somewhere warm and out of direct light, then check them every few days for contaminants. If a bag becomes contaminated, throw it out and start again. Once you see about 50% of the substrate colonized by mycelium, use your hands to press on the substrate and break up the colonized sections. Your goal here is to redistribute the mycelium throughout the bag in order to aid colonization. Break the colonized substrate, knead it through the bag, and then gently shake to mix it throughout. Then comes another period of waiting. In a few weeks, you should have full mycelial colonization.

Fruiting

You can go about entering the fruiting phase one of a few ways. The easiest method is to simply open the bag and place it in a better lit area. The presence of extra air and light will trigger the fruiting phase, and the mushrooms will start to grow from the top of the substrate block. You can also add sterilized soil to the top, following the casing technique from the McKenna method. Another option is to leave the bags sealed and use a razor blade to cut X's into the bags at different locations. Mushrooms will grow through the cuts. With both of these techniques, watering with a mist from a spray bottle will be required at about the same rate as the PF-Tek method to ensure that the substrate does not dry out.

Your other options involve removing the substrate cake entirely from the bag and treating it in the same way as the PF-Tek brown rice cakes; first soaking it, then rolling it in vermiculite, and finally placing it into a prepared fruiting chamber. This will give you a large fruiting cake that you can reuse multiple times as in the PF-Tek method. Alternatively, you can break it into pieces and mix it through a bulk substrate in a fruiting chamber, then apply casing soil, regular water, and light to achieve a large surface area of fruitbodies.

Throughout each step of all these techniques, it's important to stay sanitized, follow good sterilization measures, and keep track of what you do and when you do it. With practice and patience, plentiful psilocybin can be grown and enjoyed at home.

HOW TO TAKE A TRIP DOWN THE RABBIT HOLE

If you've read the whole book up to this point, you know just about everything there is to know about magic mushrooms: their history, their cultural significance, their chemistry, and how to grow them. What we haven't yet touched on is how exactly you're supposed to have a mushroom experience, so that will be the penultimate chapter of this book. Having a mushroom trip can be one of the most significant experiences of a person's life, and we recommend it wholeheartedly to nearly everyone. There are a couple things you can know beforehand to ensure you have the best trip possible, some of which we will give you from personal experience, while other information in this section comes from the mushroom masterminds of the last few decades. If you're ready to take a trip to Wonderland, this chapter is for you.

THE RISKS INVOLVED

Gossip about psychedelics is full of horror stories. There are anecdotal reports of all sorts- tales of single-psilocybin doses resulting in a lifelong trip, unfortunate hippies dropping acid and totally losing the remainder of their sanity, MDMA melting literal holes in your brain. One of the author's was once told of a guy who took acid and spent the rest of his life sleeping standing up; the poor guy was stuck in a trip in which he thought he believed himself to be a glass of milk and that lying down would spell his demise.

These are nothing but myths and fairy tales. There is no scientific literature supporting anything of this sort, and there's actually substantial evidence to the contrary. That is, the effects of psychedelics, even their worst outcomes, rarely persist for more than a day, with an estimated 0.18% of psychedelic users experiencing psychotic symptoms after a forty-eight-hour period.[117]

With LSD use comes a possible risk of a condition known as Hallucinogen Persisting Perception Disorder (HPPD), but even this has been called into question, and the facts of the diagnoses of this disorder remain doubtful. HPPD is said to be caused by psychoactive drug use and result in trip "flashbacks" involving perceptual distortion, feelings of intoxication, and other hallucinations. As it is now, studies involving psilocybin users and regular participants in peyote ceremonies show no prevalence of HPPD.[118] In fact, most patients who complain of hallucinogen-like perceptual "flashbacks" have never used psychedelics, and the correlation between HPPD and psychedelic use is altogether insubstantial.

The major risk you may read or hear about most often is that psychedelic use can trigger psychotic disorders such as schizophrenia. People are often warned that if they have had previous mental health diagnoses or have a family history of psychosis, they should avoid hallucinogenic drugs. However, there is no medical literature supporting this claim. Recent studies have found that lifetime psychedelic users are actually less likely to have been admitted for inpatient psychiatric treatment than those who had never used psychoactives.[119]

The conclusion of all the leading researchers in this field is that psychedelics are overall safe for most people and are not linked to an increased risk of mental health issues.[120] Of course, if you feel that taking psilocybin will be a bad idea for you, don't do it. But, the risk of tripping for the rest of your life or coming to believe that you are a glass of milk 'til you die is basically nonexistent. "Bad" trips are always a possibility, but there are some steps you can take to increase your chances of having a positive experience.

BEST PRACTICES

There are no hard and fast rules for how to take mushrooms the right way, but we do have about seventy years' worth of advice from an eclectic assortment of psychonauts that can give you some good guidelines.

How to Prepare for Your First Mushroom Experience

Many people will prepare for a trip with an assortment of activities they think they will do once the drug kicks in. They might advise you to have paper and paint and crayons for making art, snacks to eat throughout the trip, the perfect playlist ready to roll, or any other collection of trip-enhancing odds and ends. Whatever you decide to prepare, you need to make sure you have a day free to dedicate to the trip itself and the comedown. Psilocybin trips can last up to nine hours, and afterward, it is advised to rest and give yourself time to integrate the experience.

In the authors' experience, these preparations are mostly in vain. In our first mushroom experiences, we did indeed have coloring materials at the ready, along with an array of fruits and nuts and granola. We had notebooks and pens in case writing inspiration should strike, and of course, we had incense to burn to set the mood. We used exactly none of it every time. Once the trip starts, the deepest part of our consciousness is always pretty clear on its desire to simply be still and observe. Not once have we spent a mushroom trip in this way of quiet observation and come out disappointed on the other side. As a matter of fact, we believe it is the best way to trip. Eliminate distractions, be free and untasked, and simply lie down and listen.

You may still wish to prepare some sort of activities for your own trip; to each their own. The authors know people who trip and paint with great results, as well as those who get a nearly unquenchable craving for grapes or the like. It can't really hurt anything to have a little bit of a plan, but be sure not to stress yourself out mid-experience that you haven't touched a brush or that your snacks are losing their freshness. Above all, enjoy the ride.

We do heavily advise making sure you have two things prepared. The first is water. Magic mushrooms seem to bring us strongly back to the awareness that we are biological animals, and oftentimes the need for water becomes our most compelling thought. Have some drinking water nearby to avoid any unnecessary journeys outside your predetermined tripping space. The second advised preparation is to ensure you have easy access to a place to pass this water. We recommend not tripping in a place where you can't find a toilet. The initial stomach upset that accompanies many mushroom experiences will probably make you want to take a precautionary trip to the head. And, take it from us, there's little worse than being 4 hours deep in a mushroom journey and not having a place to pee. Of all the small things that can start the kind of negative thought-loop that will lead into a bad trip, needing to pass water and having no place to do so is among the most common.

Time-Honored Advice: Set and Setting

"Set and setting" was first described by Timothy Leary in his 1964 psychedelic guide, *The Psychedelic Experience: A Manual Based on the Tibetan Book of the Dead.*[121] It has since become some of the most succinct advice on how to ensure you have the best psilocybin experience possible.

Setting is the easiest of these two criteria to dial in. It simply refers to where you trip and what that place is like. Opinions vary on what is the best setting for a magic mushroom trip, from the Mazatec belief that they should only be taken inside at night in complete darkness, to those who think that shrooms are a great drug for dancing the night away at a rave. The authors prefer to trip in one of two ways.

We lend a lot of credence to the spiritual power of mushrooms and believe that a lot of wisdom can be imparted on the user if one is willing to listen. This is why we think that the traditional setting of still darkness is perhaps the most effective way to have a life-enhancing experience. Removing distractions and placing yourself in a setting that is conducive to meditative thought and self-exploration allows the voice of the mushrooms to speak loud and clear.

Our second preferred way to trip is in nature, especially in forests. As humans, we have done about as much as we can to remove ourselves from our animal origins, though hairless primates we remain. Taking mushrooms outdoors in a natural area provokes a reconnection with the environment and a doorway to our primal past. When we have a trip in the forest, it is as if the Earth begins to speak, and the vibrant vitality of our planet, along with the need to preserve our natural world, becomes incredibly apparent.

We won't tell you that tripping at a party or any other way is exactly wrong, but we highly recommend giving mushrooms the chance to be the center of attention in your first trips. They can be the most effective spirit medicine for those who seek a salve and who are willing to open their minds to the possibility that these gentle fungi hold a great power to advance each user's consciousness.

Set can be a bit more difficult to define in precise terms. The word itself is short for "mindset," and refers to the mental-emotional state of the user before and during the psychedelic experience. In short, set is your attitude in regards to the trip. Your mental-emotional state when going into a magic mushroom experience will have a huge impact on the results of the psilocybin. This drug tends to amplify those emotions that are already present, acting as a magnifying glass to our thoughts and feelings.

Perhaps the most important aspect of your mindset you should focus on is having an open mind about the coming experience. Go into the trip with an attitude of curiosity and innocence, without imposing restrictions on what you may experience. Allow the mushrooms to do what they will and to show you what they have to show. Don't limit yourself with predetermined expectations or fears, and be open to new possibilities and new realities.

Starting a trip in a negative state of mind is usually not a great idea. Though current clinical trials are studying the effects of psilocybin in those with chronic depression, these tests are being overseen by trained therapists who can help to guide the users, and who also ensure that the setting remains non-threatening and peaceful. If you are feeling extremely depressed, angry, or otherwise down, a trip may only serve to expand these feelings without the proper guidance of an outside helper. Wait until the right time and mindset, or otherwise take steps to put yourself in a better state. You may be able to achieve a more appropriate set by preparing beforehand. Do these things before eating the mushrooms.

The first thing you can do is to connect with your body and the reality of your being. Practice bodily awareness and deep breathing. Focus on the feeling of life in your body. Bring your awareness to your feet, wiggle your toes, and direct your consciousness to the feeling of aliveness there. Slowly and mindfully carry your attention up your legs, staying aware of yourself and your existence. Stay centered in the reality of your being alive, continuing to breathe deeply and fully, and lead your consciousness past your legs, into your torso, through your arms, and gradually to the top of your head. This is a simple meditative practice that is meant to give you a sense of grounding.

Do what you must to relax. Listen to music that calms you and heightens your mood. Maybe take a soothing bath. Go for a relaxing walk. The point of this is to allow yourself to be open-minded and ready to surrender to the oncoming trip. Remind yourself of the temporary aspect of time and experience, and allow negative thoughts to slip away. In the words of Leary (and the Beatles), "turn off your mind, relax, and float downstream."

Finally, you may wish to set an intention for your trip. It is difficult to describe how mushrooms can seem to speak to the user, but it is a distinct possibility that you will find the answer to a question you ask if you allow yourself to be open, curious, and accepting of the experience. Your intention, or goal of the trip, can be anything you wish. You may go into the trip with a rather profound question about the meaning of life or the underlying cause of one of your problems. Otherwise, you may have the intention to simply have a good time. The other option is to set the intention of no intention, or rather, ask that the mushrooms do with you what they will. The authors have utilized all three types of intentions in our trips and have had rewarding experiences each time.

Trusting in a Trip Sitter

For your first trip, or maybe first few trips, you may want to have a person with you to offer help if you start to experience a difficult time. These people are colloquially called "trip sitters". A trip sitter's purpose is to offer care and comfort, both actively and passively. Mushroom experiences can be extremely powerful and sometimes frightening, so having a person to accompany you during the experience can help to ease any fears or issues that may arise.

A trip sitter should be someone that you trust and are comfortable around. Ideally, they should be sober during your experience and willing to remain with you for the duration of the trip. Trip sitters should be familiar with the effects of mushrooms and know in what capacity they will be able to offer assistance. Compassion is a huge part of being a sitter, as well as knowing when and when not to interfere with the user.

When taking mushrooms, there is always the possibility to experience a "bad" trip. You may find yourself in a type of recursive thought loop in which your mind becomes focused on a particular idea. If this idea is negative, continually returning to the thought can start to manifest feelings of fear, anxiety, and panic. A trip sitter will be able to bring your attention to something else and help you to break this cycle of thought.

Additionally, a trip sitter should care for your physical well-being. This includes helping you to drink water, move when needed, find and perhaps use the bathroom, and keep you from harming yourself either accidentally or intentionally. Being a trip sitter is a role of responsibility, and should only be entrusted to those who you know have your best intentions in mind.

After you become more familiar with psilocybin, you may no longer wish for the presence of or require an additional person to be there with you. It is never a bad idea to trip with trusted company, but once you are more aware of the potential of a magic mushroom experience, it should be fine to trip alone.

Know Your Drug, Know Your Dose

The final thing you should consider before you trip is how much you will take. This is largely a matter of preference, but it's important to know what you can expect at different dosages. For first time users, especially if you are going into the experience without a knowledgeable sitter, we would recommend starting with a low dose. If you acquaint yourself with the power of mushrooms with a first-time small amount, you will have an idea of what they are capable of and what you can expect going forward.

With that said, if you have someone to guide you in case things go awry, there is nothing wrong with taking a large dose your first time. If it is your very first psychedelic experience, you may experience a bit of anxiety and fear at some stages, but as long as there is someone to assure that you don't fall into physical harm, taking a deep dive into the world of mushrooms isn't dangerous in itself.

It is probably best to be somewhat aware of the effects of mushrooms so that you are not taken by complete surprise your first time in. Being a bit familiar with both their psychological and somatic effects can prevent some of the feelings of anxiety that might arise once the psilocybin kicks in. If you've read the whole book up to this point, you have some idea of those effects, but we'll go into it a bit further in the following section.

WHAT TO EXPECT AT VARIOUS DOSAGE LEVELS

There are mainly five different dosage levels we can categorize our mushroom experiences into: microdosing, low, medium, high, and "heroic" dosing. The effects are exponentially different between the micro and heroic levels, and there's not much in the way of comparison between the two. Of important note is that dosage levels in dried grams of mushrooms will not produce the same effects for everyone. There are various factors that affect how you will respond to different amounts of mushrooms, such as your body weight, your metabolism, your mindset, other medications or drugs you are using, and whether or not you eat before the experience. Keep these things in mind when choosing your dose. If you don't want a particularly strong trip, the best advice is to start with less than you think you'll need, and gradually increase the dose in your following trips.[124]

Microdosing

Microdosing is defined as about one-tenth the amount of a hallucinogenic amount of a psychedelic. For psilocybin, this comes out to around a third of a gram of dried mushrooms. Microdosing generally won't cause any strong changes in perception, and a recent study in the Netherlands showed that this amount of mushrooms doesn't noticeably affect rational thinking, abstract reasoning, or problem-solving abilities.[122] However, this microdose level of psilocybin can increase creativity, improving convergent and divergent thinking.

People taking mushroom microdoses often claim that their senses are heightened, including increased visual and auditory acuity and higher energy levels. Further purported benefits include a lessening of anxiety, a sharper consciousness, more attuned creative thinking, and reduced levels of depression. This is also considered to be an effective dose for relieving cluster headaches and migraines.[123]

Microdosing can be done at more frequent intervals than hallucinogenic doses. Tolerance to psilocybin develops rather quickly, in that if you take a large dose one day, you will have to take much more to have the same effects the next day. It is typically recommended to wait a week between hallucinogenic-level trips, but microdosers often shorten this gap to two or three days between doses.

Low Dose

A low dose of psilocybin can range from anywhere between a half gram to one and a half grams of dried mushrooms, largely dependent on your body weight. Some users may experience strong effects at even one gram, though this is a rare occurrence, while others may barely notice any perceivable effects at the same dose.

Effects will start roughly between twenty to sixty minutes after swallowing the mushrooms. Here, we must advise you against "stacking" doses. If you don't feel anything after an hour of eating the mushrooms, don't take another dose just yet. Doing this is called "stacking" and can result in much stronger than desired effects once the trip starts.

The negative reactions to psilocybin at low doses are generally limited to feelings of anxiety or nervousness, perhaps more linked to users' anticipation than anything else. These typically melt away with the onset of the more positive results, which commonly begin with changes in visual perception. You may notice that lights and colors appear to be brighter, and lights may develop a glimmer or sparkle similar to stars. Visual trails can occur, where you will see a faint sort of afterimage following moving objects.

Shortly after the visual effects begin, you will probably experience a feeling of giddiness or seemingly causeless happiness. From here, there is often a heightening in both emotions and awareness, together with feelings of bodily lightness and bliss.

Medium Doses

The medium dose of magic mushrooms is to many users the most pleasant and preferable. In general, a medium dose is less than two and one-half grams of dried mushrooms. In addition to all the low dose effects, medium doses come with more pronounced changes in visual perception. Lights may develop a "rainbow" component, seeming to shine with prismatic effulgence. Here, the so-called "hallucinations" can begin. Users may see distinct shapes and forms with eyes either opened or closed. What's more, is the change in temporal perception. During a mushroom experience, your perception of time can become greatly distorted, such that a second may seem to take minutes to pass, minutes hours, and hours days or even eons.

You may be swept up by a sense of euphoria or "rightness." The influence on thoughts at medium doses can be quite extraordinary. At this level, users frequently report feelings of realization about past and current situations, new plateaus of emotional awareness, a consciousness of the interconnectivity of reality, and other general and specific epiphanies.

There can be rather contradictory effects as well. You may have an increased ability to focus or a propensity for distraction. There might be a feeling of great joy and connectedness or feelings of isolation and despair. Much of this is dependent on the set and setting of the experience, as well as the user's mental-emotional state prior to and during the episode.

It is at this level that many users report feelings of well-being, great happiness and contentment, and personal insights regarding themselves and the workings of the universe. Religious or spiritual experiences are likely with medium doses.

However, not every trip is positive and pleasant, and we now offer you a few words of advice and caution. Our intent is not to scare you away from what is by and large a predominantly beneficial ordeal, but simply to arm you with foreknowledge that may prove useful if you chance to have a "bad trip." Psychedelic experiences can sometimes be very difficult, with both mental and emotional suffering. Intense feelings of anxiety and despair may arise. One commonly reported negative effect is the feeling that the trip will never end; that the user will be stuck in this semi-sane state forever. If this should happen to you, try to remind yourself that what you have done is eaten a mushroom whose effects are inarguably temporary and that whatever negative feelings you are experiencing in the depth of the trip will pass after a matter of hours. As magic mushrooms so often teach, this too shall pass.

High Doses

High doses of two and one-half grams or more are among the most interesting, unexplored, and undocumented aspects of humanity. Upon ingestion of these amounts, users routinely report life-altering realizations and epiphanies that may influence their behavior well after the trip has passed.

There can be a sense of loss of self, including ego death, a sense of becoming one with all reality, and even experiencing the feeling of dying. Seemingly autonomous beings may appear to the user and impart wisdom and advice. With this comes intense feelings of amazement, wonder, joy, and awe. Time dilation becomes extreme, and all perception of time may cease so that trips can seem to exist outside of the temporal dimension or take thousands of years to pass. Feelings of righteous elation and euphoria are common, along with a feeling of awakening from a previous sleep one has been in up to this point. These high dose trips may have an "afterglow" effect in which a feeling of bliss and well-being extends for several days after the mushroom experience.

The authors have had intense religious experiences at high doses, having the feeling that we are viewing the heavenly dimension or directly communing with God. These experiences, though temporary, have stayed with us for years and in large part formed the basis of our overall senses of peace and contentment in life.

The "Heroic" Dose

Terence McKenna was known to recommend a dose of five grams to feel full effects, dubbing this amount the "heroic dose." Advising users to take this amount alone in a dark room, McKenna claims this as the threshold dose at which guiding voices may truly speak to the users, stating "five dried grams will flatten even the most resilient ego".

The McKenna brothers posited that magic mushrooms may be of extraterrestrial origin, with either the mushrooms being aliens themselves or a type of communication device used by a more consciously-advanced alien society to communicate with other beings. At doses of five grams, it does often happen that a seemingly "outside" voice will communicate with the user. In one of the author's experience, this voice takes the form of a kindly, somewhat amused, loving presence that offers insight into his own flaws in thinking and various poor mental habits. It isn't condescending, but rather pokes fun at the ego and the significance given to miniscule human thoughts and beliefs.

This "alien presence" sometimes seems like a big brother, offering advice as to how to be at peace with oneself and with the world. It imparts revelations that are, at the time of the trip, completely life-shattering. However, the form of communication does not translate into normal consciousness, and during the comedown phase of the trip, many seem to forget all of the most incredible insights.

The dissolution of the ego can be a frightening experience and is wholly beyond words. It is essentially a melting-away of your concept of individual self, and a sense of becoming an equal, inseparable part of all reality. It can feel like dying, and indeed is a temporary form of the death of your individuality. You may feel the urge to fight against this, as is natural, but we advise you to surrender to the guidance of the psilocybin and experience the bliss of becoming nothing. Do as the fruiting bodies of the mushrooms do, and allow yourself to fade away, back into the all-encompassing web of the interconnecting secret mycelium that lies just beneath the surface of consciousness. In the words of the poet Rumi:

> The candle as it diminishes explains,
> "Gathering more and more is not the way.
> Burn, become light and heat and help.
> Melt."

CLOSING

With this, we end our look into the world of psilocybin. There is much we still do not know, and much that probably remains unknowable, but hopefully we will continue to learn more about this special group of species as time goes on. Though we may never understand the full extent of their historical significance and the role they played in shaping humans and society over the course of time, we can practice respectful and responsible use going forward to ensure that future generations will look upon these magic mushrooms with a meaningful and significant recognition of their benefits.

In your dealings with psilocybin mushrooms and all other psychedelics, approach with an open mind, yet equip yourself with the knowledge and understanding that is available to us in the modern age. Practice responsible drug use, being respectful of the power of these psychoactive substances, and doing what you can to remove the taboo associated with their use. With a proactive, mindful approach, you can gain an invaluable wealth of wisdom, insight, and healing from the sacred fungi.

And so ends our ode to the magic mushroom. It is our wish that this book brings you new knowledge, expands your curiosity, and does its part in dispelling the stigma that has befallen psilocybin in the last century. We look with a hopeful heart toward the future and believe that in due time, magic mushrooms will find a place of prominence in our society. The little saints may be the healers we need to find collective peace and prosperity as we move forward as a species.

RECOMMENDED READING

Barring the authors' own experiences with psilocybin, we owe everything in this book to writers and researchers before us. Though the following list is far from exhaustive, the books in this recommended reading section can provide you with a deeper understanding and appreciation of magic mushrooms and other psychedelics, extending your knowledge of their history, their chemistry, and their significance in many dimensions.

Reading List

Arora, David. *Mushrooms Demystified.* 1979.

Fadiman, James. *The Psychedelic Explorer's Guide: Safe, Therapeutic, and Sacred Journeys.* 2011.

Furst, Peter. *The Encyclopedia of Psychoactive Drugs: Mushrooms.* 1986.

Leary T, Metzner R, Alpert R. *The Psychedelic Experience: A Manual Based on the Tibetan Book of the Dead.* 1964.

Letcher, Andy. *Shroom: The Cultural History of the Magic Mushroom.* 2006.

Mandrake K, Haze V. *The Psilocybin Mushroom Bible: The Definitive Guide to Growing and Using Magic Mushrooms.* 2016.

McKenna, Terence. *Food of the Gods: The Search for the Original Tree of Knowledge.* 1992.

McKenna, Terence. *True Hallucinations.* 1989.

Narby, Jeremy. *The Cosmic Serpent: DNA and the Origins of Knowledge.* 1998.

Oss O.T., Oeric O.T. *Psilocybin: Magic Mushroom Grower's Guide.* 1986.

Stamets, Paul. *Psilocybin Mushrooms of the World.* 1996.

Pollan, Michael. *How to Change Your Mind.* 2018.

Powell, Simon G. *The Psilocybin Solution: The Role of Sacred Mushrooms in the Quest for Meaning.* 2011.

Wasson, R. Gordon. *Soma: Divine Mushroom of Immortality.* 1968.

REFERENCES LIST

1. Guzmán, Gastón & Allen, John & Gartz, Jochen. (1998). A Worldwide geographical distribution of the Neurotropic Fungi, an analysis and discussion. Ann Mus Civ Rovereto. 14.

2. Guzmán, Gastón. (2005). "Species diversity of the genus *Psilocybe* (Basidiomycotina, Agaricales, Strophariaceae) in the world mycobiota, with special attention to hallucinogenic properties". *International Journal of Medicinal Mushrooms*. **7** (1–2): 305–331.

3. NIDA (1969). Hallucinogens. Retrieved April 19, 2020, from https://www.drugabuse.gov/publications/drugfacts/hallucinogens.

4. Bleyer J. (May 2017). A Radical New Approach to Beating Addiction. *Psychology Today*.

5. Cerletti, A. U. R. É. L. I. O. (1958). Etude pharmacologique de la psilocybine. *Les champignons hallucinogenes du mexique. Paris: Museum de historie naturelle*, 268-71.

6. Lee, Martin A. (1985). *Acid Dreams: The Complete Social History of LSD: The CIA, The Sixties, and Beyond*. Grove Press. 39.

7. Erowid. (1997, February 2). Psilocybin Mushrooms Effects. Retrieved April 20, 2020, from https://erowid.org/plants/mushrooms/mushrooms_effects.shtml.

8. Erowid. (1997, February 2). Psilocybin Mushrooms Effects. Retrieved April 20, 2020, from https://erowid.org/plants/mushrooms/mushrooms_effects.shtml.

9.   Sahagûn, B. (2011). *Historia General de las Cosas de Nueva España* (Cambridge Library Collection - Latin American Studies) (C. Bustamante, Ed.). Cambridge: Cambridge University Press.

10. Schultes, R. E. (1940). Teonanacatl: the narcotic mushroom of the Aztecs. *American Anthropologist, 42*(3), 429–443. doi: 10.1525/aa.1940.42.3.02a00040.

11. Pettigrew, J. (2011), Iconography in Bradshaw rock art: breaking the circularity. Clinical and Experimental Optometry, 94: 403-417. doi:10.1111/j.1444-0938.2011.00648.x.

12. Samorini, G. (2012). Mushroom effigies in world archaeology: from rock art to mushroom-stones. *The stone mushrooms of Thrace, EKATAIOS, Alexandroupoli*, 16-44.

13. Akers, B. P., Ruiz, J. F., Piper, A., & Ruck, C. A. (2011). A prehistoric mural in Spain depicting neurotropic Psilocybe mushrooms?. *Economic Botany, 65*(2), 121-128.

14. Mayer, K. H. (1977). *The mushroom stones of Mesoamerica*. Acoma Books.

15. Hernández-Santiago, F., Martínez-Reyes, M., Pérez-Moreno, J., & Mata, G. (2017). Pictographic representation of the first dawn and its association with entheogenic mushrooms in a 16th century Mixtec Mesoamerican Codex. *Revista Mexicana de Micología, 46*, 19-28.

16. Guzmán, G. (2012). New taxonomical and ethnomycological observations on Psilocybe SS (fungi, Basidiomycota, agaricomycetidae, Agaricales, Strophariaceae) from Mexico, Africa and Spain. *Acta Botanica Mexicana*, (100), 79-106.

17. Schultes, R. E. (1940). Teonanacatl: the narcotic mushroom of the Aztecs. *American Anthropologist, 42*(3), 429-443.

18. Kamieński Dłużyk, A. Diariusz więzienia moskiewskiego, miast i miejsc [в:] A. Kamieński Dłużyk. *Dwa polskie pamiętniki z Syberii. XVII i XVIII wiek*, 382.

19. Jochelson, W. (1905). The Koryak. Memoirs of the AMNH; v. 10, pt. 1-2; Publications of the Jesup North Pacific Expedition; v. 6. 120-121.

20. Enderli, J. (1903). Zwei Jahre bei den Tschuktschen und Korjaken. *Petermanns Geographische Mitteilungen, 49*(8), 183-185.

21. Karjalainen, K. F., & Hackman, O. (1922). *Die religion der Jugra-völker* (Vol. 1). Suomalainen tiedeakatemia. 278-280.

22. Saar, M. (1991). Ethnomycological data from Siberia and North-East Asia on the effect of Amanita muscaria. *Journal of ethnopharmacology, 31*(2), 157-173.

23. Von Strahlenberg, P. J. (1738). *An Historico-geographical Description of the North and Eastern Parts of Europe and Asia: But More Particularly of Russia, Siberia, and Great Tartary; Both in Their Ancient and Modern State: Together with an Entire New Polyglot-table of the Dialects of 32 Tartarian Nations*. W. Innys and R. Manby. 396.

24. Lutkajtis, A. (2020). Lost Saints: Desacralization, Spiritual Abuse and Magic Mushrooms. *Fieldwork in Religion, 14*(2), 118-139.

25. Sabina, M., & Estrada, A. (2003). *María Sabina: Selections* (Vol. 2). Univ of California Press.

26. Stevens, J. (1998). *Storming heaven: LSD and the American dream*. Grove Press. 326.

27. Leary, T. (1990). *Flashbacks: a personal and cultural history of an era: an autobiography*. Tarcher. 253.

28. Melton, J. G. (2016, April 7). Realizing the New Age. Retrieved April 24, 2020, from https://www.britannica.com/topic/New-Age-movement/Realizing-the-New-Age

29. Kent, James L. (2010). *Psychedelic Information Theory: Shamanism in the Age of Reason*, Appendix, 'Why are Psychedelics Spiritual?'. PIT Press, Seattle.

30. Capasso, L. (1998). 5300 years ago, the Ice Man used natural laxatives and antibiotics. *The Lancet*, *352*(9143), 1864.

31. Molitoris, H. P. (1994). Mushrooms in medicine. *Folia microbiologica*, *39*(2), 91.

32. Shao, L. I., & ZHANG, B. (2013). Traditional Chinese medicine network pharmacology: theory, methodology and application. *Chinese journal of natural medicines*, *11*(2), 110-120.

33. Welz, A. N., Emberger-Klein, A., & Menrad, K. (2019). The importance of herbal medicine use in the German health-care system: prevalence, usage pattern, and influencing factors. *BMC Health Services Research*, *19*(1), 952.

34. Oyebode, O., Kandala, N. B., Chilton, P. J., & Lilford, R. J. (2016). Use of traditional medicine in middle-income countries: a WHO-SAGE study. *Health policy and planning*, *31*(8), 984-991.

35. Kala, & Kala, Chandra Prakash & Sajwan, Bikram. (2007). Revitalizing Indian system of herbal medicine by the National Medicinal Plants Board through institutional networking and capacity building. Current science. 93. 797-806.

36. Rashrash, M., Schommer, J. C., & Brown, L. M. (2017). Prevalence and Predictors of Herbal Medicine Use Among

Adults in the United States. *Journal of patient experience, 4*(3), 108–113. https://doi.org/10.1177/2374373517706612.

37. Oyebode, O., Kandala, N. B., Chilton, P. J., & Lilford, R. J. (2016). Use of traditional medicine in middle-income countries: a WHO-SAGE study. *Health policy and planning, 31*(8), 984-991.

38. Farnsworth, N. R. (1988). Screening plants for new medicines. *Biodiversity, 15*(3), 81-99.

39. Sarris, J., Kavanagh, D. J., Byrne, G., Bone, K. M., Adams, J., & Deed, G. (2009). The Kava Anxiety Depression Spectrum Study (KADSS): a randomized, placebo-controlled crossover trial using an aqueous extract of Piper methysticum. *Psychopharmacology, 205*(3), 399-407.

40. Klemow, K. M., Bartlow, A., Crawford, J., Kocher, N., Shah, J., & Ritsick, M. (2011). 11 Medical Attributes of St. John's Wort (Hypericum perforatum). *Lester Packer, Ph. D.,* 211.

41. McKenna, T. (1992, October). *Earth Trust Benefit. Earth Trust Benefit.* Los Angeles, California.

42. McKenna, T. (1999). *Food of the gods: the search for the original tree of knowledge: a radical history of plants, drugs and human evolution.* Random House. 55-60.

43. Samorini, G. (1992). The oldest representations of hallucinogenic mushrooms in the world (Sahara Desert, 9000-7000 BP). *Integration, 2*(3), 69-78.

44. Akers, Brian & Ruiz, Juan & Piper, Alan & Ruck, Carl. (2011). A Prehistoric Mural in Spain Depicting Neurotropic Psilocybe Mushrooms?1. Economic Botany. 65. 121-128. 10.1007/s12231-011-9152-5.

45. Dikov, N. N., & Bland, R. L. (1999). *Mysteries in the Rocks of Ancient Chukotka: Petroglyphs of Pegtymel'*. US Department of the Interior, National Park Service, Shared Beringian Heritage Program. 39-53.

46. Dikov, N. N., & Bland, R. L. (1999). *Mysteries in the Rocks of Ancient Chukotka: Petroglyphs of Pegtymel'*. US Department of the Interior, National Park Service, Shared Beringian Heritage Program. 56-57.

47. Griffith, R. T. H. (1896). Rig-Veda. 8.48.3. *Rig Veda, tr. by Ralph TH Griffith*.

48. Wasson, R. G. (1971). Soma: Divine mushroom of immortality.

49. McKenna, T. (1999). *Food of the gods: the search for the original tree of knowledge: a radical history of plants, drugs and human evolution*. Random House, 97-120.

50. Sarianidi, V. I. (2003). Margiana and soma-haoma. *Electronic Journal of Vedic Studies*, *9*(1), 53-73.

51. Lowy, B. (1972). Mushroom symbolism in Maya codices. *Mycologia*, *64*(4), 816-821.

52. Schultes, R. E. (1940). Teonanacatl: the narcotic mushroom of the Aztecs. *American Anthropologist*, *42*(3), 429.

53. Kamieński Dłużyk, A. Diariusz więzienia moskiewskiego, miast i miejsc [в:] A. Kamieński Dłużyk. *Dwa polskie pamiętniki z Syberii. XVII i XVIII wiek*, 382.

54. Von Strahlenberg, P. J. (1738). *An Historico-geographical Description of the North and Eastern Parts of Europe and Asia: But More Particularly of Russia, Siberia, and Great Tartary; Both in Their Ancient and Modern State: Together with an Entire New Polyglot-table of the Dialects of 32 Tartarian Nations*. W. Innys and R. Manby, 397.

55. Ström, Å. (1982). Berserker und Erzbischof: Bedeutung und Entwicklung des altnordischen Berserkerbegriffes. *Scripta Instituti Donneriani Aboensis, 11*, 178-186.

56. Peschel, K. (1998). *Puhpohwee for the people: A Narrative Account of Some Uses of Fungi among the Ahnishinaabeg.* LEPS Press.

57. Hofmann, A., & Rätsch, C. (2001). *Plants of the gods: their sacred, healing, and hallucinogenic powers.* Healing Arts Press, 82-85.

58. Wasson, R. G. (1980). *The wondrous mushroom: mycolatry in Mesoamerica* (No. 7). New York: McGraw-Hill, 43-44.

59. Hillebrand, J., Olszewski, D., & Sedefov, R. (2000). *Hallucinogenic mushrooms: an emerging trend case study.* EMCDDA.

60. Brande E. (1799). "Mr. E. Brande, on a poisonous species of Agaric". *The Medical and Physical Journal: Containing the Earliest Information on Subjects of Medicine, Surgery, Pharmacy, Chemistry and Natural History.* **3**: 41–44.

61. Verrill, A. E. (1914). A recent case of mushroom intoxication. *Science, 40*(1029), 408-410.

62. Johnson, J. B. (1940). Note on the discovery of teonanacatl. *American Anthropologist, 42*(3), 549-550.

63. Schultes, R. E. (1940). Teonanacatl: the narcotic mushroom of the Aztecs. *American Anthropologist, 42*(3), 429-443.

64. Wasson, R. G. (1957). Seeking the magic mushroom. *Life, 42*(19), 100-120.

65. Estrada, A. (1981). *Maria Sabina, Her Life and Chants* (Vol. 1). Ross Erikson, 79-80.

66. Marks, J. (1979). *The search for the" Manchurian candidate": The CIA and mind control.* New York: Times Books, 115-116.

67. Facts, P. F. (2006). January 2006. *National Drug Intelligence Center.*

68. Moreno, F. A., Wiegand, C. B., Taitano, E. K., & Delgado, P. L. (2006). Safety, tolerability, and efficacy of psilocybin in 9 patients with obsessive-compulsive disorder. *Journal of Clinical Psychiatry, 67*(11), 1735-1740.

69. Pathways, C. (2018). Compass pathways receives FDA breakthrough therapy designation for psilocybin therapy for treatment-resistant depression. *CompassPathways.com.*

70. Denis-Lalonde, D., & EStefan, A. (2020). Emerging Psychedelic-Assisted Therapies. *Journal of Mental Health and Addiction Nursing, 4*(1), e1-e13.

71. Kaur, Harmeet (30 January 2020). "Santa Cruz decriminalizes magic mushrooms and other natural psychedelics, making it the third US city to take such a step". *CNN*. Retrieved 26 April 2020.

72. Wasson, R. G. (1957). Seeking the magic mushroom. *Life, 42*(19), 100-120.

73. Marks, J. (1979). *The search for the "Manchurian candidate": The CIA and mind control.* New York: Times Books, 114-117.

74. Wasson, V. P. (1957). *I ate the sacred mushrooms.* United Newspapers Magazine Corporation, 8-10, 36.

75. Weil, A. T. (1963). The strange case of the Harvard drug scandal. *Look, 27*(22), 46.

76. Doblin, R. (1998). Dr. Leary's Concord Prison Experiment: a 34-year follow-up study. *Journal of psychoactive drugs, 30*(4), 419-426.

77. ahnke, W. N. (1963). *Drugs and mysticism: An analysis of the relationship between psychedelic drugs and the mystical consciousness: A thesis* (Doctoral dissertation, Harvard University).

78. Doblin, R. (1991). Pahnke's "Good Friday experiment": A long-term follow-up and methodological critique. *Journal of Transpersonal Psychology, 23*(1), 1-28.

79. Weil, A. T. (1963). The strange case of the Harvard drug scandal. *Look, 27*(22), 46.

80. Cohen, J. E. (1963, January 16). I.F.I.F. Group Plans Center For Research. *The Harvard Crimson*. Retrieved from https://www.thecrimson.com/article/1963/1/16/ifif-group-plans-center-for-research/.

81. Lee, M. A., & Shlain, B. (1992). *Acid dreams: The complete social history of LSD: The CIA, the sixties, and beyond.* Grove Press, 92-93.

82. United Nations Conference for the Adoption of a Protocol on Psychotropic Substances (1971: Vienna, Austria). (1977). *Convention on Psychotropic Substances, 1971: Including Final Act and Resolutions, as Agreed by the 1971 United Nations Conference for the Adoption of a Protocol on Psychotropic Substances, and the Schedules Annexed to the Convention.* New York: United Nations.

83. Martindale, C., & Fischer, R. (1977). The effects of psilocybin on primary process content in language. *Confinia psychiatrica*.

84. (2017, February 2). Retrieved from https://www.youtube.com/watch?v=npXgLZuvIo4.

85. Erowid. (1999, October 2). Psilocybin Mushrooms Timeline. Retrieved April 27, 2020, from https://erowid.org/plants/mushrooms/mushrooms_timeline.php.

86. Ott, J. (1978, February). The 1978 World Conference on Hallucinogenic Mushrooms. *Head Magazine*.

87. Allen, J. W. (2002). Mushroom Pioneers: R. Gordon Wasson, Richard Evans Schultes, Albert Hofmann, Timothy Francis

Leary and Others. *Ethnomycological Journals Sacred Mushroom Studies, 7,* 198.

88. Griffiths, R. R., Richards, W. A., McCann, U., & Jesse, R. (2006). Psilocybin can occasion mystical-type experiences having substantial and sustained personal meaning and spiritual significance. *Psychopharmacology, 187*(3), 268-283.

89. Jones, H. (2019, September 4). Johns Hopkins launches center for psychedelic research. Retrieved from https://hub.jhu. edu/2019/09/04/hopkins-launches-psychedelic-center/.

90. Mission. (n.d.). Retrieved from https://maps.org/about/ mission/.

91. David E. Nichols (2014) The Heffter Research Institute: Past and Hopeful Future, Journal of Psychoactive Drugs, 46:1, 20-26, doi: 10.1080/02791072.2014.873688.

92. About the Foundation. (n.d.). Retrieved from https:// beckleyfoundation.org/about/the-foundation/.

93. Petranker, R., & Anderson, T. (2017). Global Drug Survey. Retrieved from https://www.globaldrugsurvey.com/ wp-content/themes/globaldrugsurvey/results/GDS2017_ key-findings-report_final.pdf.

94. Kirsch, I., Deacon, B. J., Huedo-Medina, T. B., Scoboria, A., Moore, T. J., & Johnson, B. T. (2008). Initial severity and antidepressant benefits: a meta-analysis of data submitted to the Food and Drug Administration. *PLoS medicine, 5*(2), e45. https://doi.org/10.1371/journal.pmed.0050045.

95. Hendricks, P. S., Thorne, C. B., Clark, C. B., Coombs, D. W., & Johnson, M. W. (2015). Classic psychedelic use is associated with reduced psychological distress and suicidality in the

United States adult population. *Journal of Psychopharmacology*, *29*(3), 280–288. doi: 10.1177/0269881114565653.

96. Carhart-Harris, R. L., Bolstridge, M., Day, C., Rucker, J., Watts, R., Erritzoe, D. E., Kaelen, M., Giribaldi, B., Bloomfield, M., Pilling, S., Rickard, J. A., Forbes, B., Feilding, A., Taylor, D., Curran, H. V., & Nutt, D. J. (2018). Psilocybin with psychological support for treatment-resistant depression: six-month follow-up. *Psychopharmacology*, *235*(2), 399–408. https://doi.org/10.1007/s00213-017-4771-x.

97. Saplakoglu, Y. (2019, November 25). FDA Calls Psychedelic Psilocybin a 'Breakthrough Therapy' for Severe Depression. Retrieved from https://www.livescience.com/psilocybin-depression-breakthrough-therapy.html.

98. How Common is PTSD in Adults? (2018, September 13). Retrieved from https://www.ptsd.va.gov/understand/common/common_adults.asp.

99. Hoskins M, Pearce J, Bethell A, et al. (2015) Pharmacotherapy for post-traumatic stress disorder: systematic review and meta-analysis. *Br J Psychiatry*. 206(2):93[]100. doi:10.1192/bjp.bp.114.148551.

100. Catlow, Briony & Song, Shijie & Paredes, Daniel & Kirstein, Cheryl & Sanchez-Ramos, Juan. (2013). Effects of psilocybin on hippocampal neurogenesis and extinction of trace fear conditioning. Experimental brain research. Experimentelle Hirnforschung. Experimentation cerebrale. 228. 10.1007/s00221-013-3579-0.

101. Carhart-Harris, R. L., Erritzoe, D., Williams, T., Stone, J. M., Reed, L. J., Colasanti, A., … Nutt, D. J. (2012). Neural correlates of the psychedelic state as determined by fMRI studies with

psilocybin. *Proceedings of the National Academy of Sciences, 109*(6), 2138–2143. doi: 10.1073/pnas.1119598109.

102. Griffiths, R. R., Johnson, M. W., Carducci, M. A., Umbricht, A., Richards, W. A., Richards, B. D., ... Klinedinst, M. A. (2016). Psilocybin produces substantial and sustained decreases in depression and anxiety in patients with life-threatening cancer: A randomized double-blind trial. *Journal of Psychopharmacology, 30*(12), 1181–1197. doi: 10.1177/0269881116675513.

103. Ross, S., Bossis, A., Guss, J., Agin-Liebes, G., Malone, T., Cohen, B., ... Schmidt, B. L. (2016). Rapid and sustained symptom reduction following psilocybin treatment for anxiety and depression in patients with life-threatening cancer: a randomized controlled trial. *Journal of Psychopharmacology, 30*(12), 1165–1180. doi: 10.1177/0269881116675512.

104. Pisano, V. D., Putnam, N. P., Kramer, H. M., Franciotti, K. J., Halpern, J. H., & Holden, S. C. (2017). The association of psychedelic use and opioid use disorders among illicit users in the United States. *Journal of Psychopharmacology, 31*(5), 606–613. doi: 10.1177/0269881117691453.

105. Bogenschutz, M. P., Forcehimes, A. A., Pommy, J. A., Wilcox, C. E., Barbosa, P., & Strassman, R. J. (2015). Psilocybin-assisted treatment for alcohol dependence: A proof-of-concept study. *Journal of Psychopharmacology, 29*(3), 289–299. doi: 10.1177/0269881114565144.

106. Johnson, M. W., Garcia-Romeu, A., Cosimano, M. P., & Griffiths, R. R. (2014). Pilot study of the 5-HT2AR agonist psilocybin in the treatment of tobacco addiction.

Journal of Psychopharmacology, 28(11), 983–992. doi: 10.1177/0269881114548296.

107. Hendricks, P. S., Clark, C. B., Johnson, M. W., Fontaine, K. R., & Cropsey, K. L. (2014). Hallucinogen use predicts reduced recidivism among substance-involved offenders under community corrections supervision. *Journal of Psychopharmacology, 28*(1), 62–66. doi: 10.1177/0269881113513851.

108. Hendricks, P. S., Crawford, M. S., Cropsey, K. L., Copes, H., Sweat, N. W., Walsh, Z., & Pavela, G. (2017). The relationships of classic psychedelic use with criminal behavior in the United States adult population. *Journal of Psychopharmacology, 32*(1), 37–48. doi: 10.1177/0269881117735685.

109. Mason, N. L., Mischler, E., Uthaug, M. V., & Kuypers, K. P. (2019). Sub-acute effects of psilocybin on empathy, creative thinking, and subjective well-being. *Journal of psychoactive drugs, 51*(2), 123-134.

110. Mariano, M., Pino, M. C., Peretti, S., Valenti, M., & Mazza, M. (2016). Understanding criminal behavior: Empathic impairment in criminal offenders. *Social Neuroscience, 12*(4), 379–385. doi: 10.1080/17470919.2016.1179670.

111. Bauer, B. (2020, January 8). The Pharmacology of Psilocybin and Psilocin. Retrieved from https://psychedelicreview.com/the-pharmacology-of-psilocybin-and-psilocin/.

112. Carhart-Harris, R. L., Erritzoe, D., Williams, T., Stone, J. M., Reed, L. J., Colasanti, A., ... Nutt, D. J. (2012). Neural correlates of the psychedelic state as determined by fMRI studies with psilocybin. *Proceedings of the National Academy of Sciences, 109*(6), 2138–2143. doi: 10.1073/pnas.1119598109.

113. Psilocybin-Facilitated Smoking Cessation. (n.d.). Retrieved April 27, 2020, from https://www.quitsmokingbaltimore.org/.

114. Host Defense. (2017, June 29). The Mushroom Life Cycle. Retrieved from https://hostdefense.com/blogs/host-defense-blog/the-mushroom-lifecycle.

115. Oss, O. T., & Oeric, O. N. (1991). *Psilocybin: magic mushroom growers guide: a handbook for psilocybin enthusiasts.* San Francisco, Calif.?: Quick American Pub.

116. McPherson, R. (n.d.). Retrieved from https://web.archive.org/web/20071011043146/http://www.fanaticus.com/.

117. Cohen, S. (1960). Lysergic acid diethylamide: side effects and complications. *J Nerv Ment Dis, 130*(Jan), 30-40.

118. Krebs, T. S., & Johansen, P. Ø. (2013). Psychedelics and mental health: a population study. *PloS one, 8*(8), e63972.

119. Johansen, P. Ø., & Krebs, T. S. (2015). Psychedelics not linked to mental health problems or suicidal behavior: A population study. *Journal of Psychopharmacology, 29*(3), 270-279.

120. Krebs, T. S., & Johansen, P. Ø. (2013). Psychedelics and mental health: a population study. *PloS one, 8*(8), e63972.

121. Alpert, R., Leary, T., & Metzner, R. (1964). The Psychedelic Experience: a Manual Based on the Tibetan Book of the Dead, 1-4.

122. Prochazkova, L., Lippelt, D. P., Colzato, L. S., Kuchar, M., Sjoerds, Z., & Hommel, B. (2018). Exploring the effect of microdosing psychedelics on creativity in an open-label natural setting. *Psychopharmacology, 235*(12), 3401-3413.

123. Andersson, M., Persson, M., & Kjellgren, A. (2017). Psychoactive substances as a last resort-a qualitative study of self-treatment of migraine and cluster headaches.

Harm reduction journal, 14(1), 60. https://doi.org/10.1186/
s12954-017-0186-6.

124. Erowid. (1997, February 2). Psilocybin Mushrooms Effects.
Retrieved April 20, 2020, from https://erowid.org/plants/
mushrooms/mushrooms_effects.shtml.

125. McKenna, T. (1991). The archaic revival. *New York:
HarperSanFrancisco*, 15.

★ ★ ★ ★ ★

Thank you for getting our book!

If you enjoy reading and find it useful we would greatly appreciate your review on Amazon.

Just head on over to this book's Amazon page and click "Write a customer review".

We read each and every one of them. Thanks!

★ ★ ★ ★ ★